Health Effects of Tea and Its Catechins

Health Effects of Tea and Its Catechins

Yukiaki Kuroda
National Institute of Genetics
Shizuoka, Japan

and

Yukihiko Hara
Tokyo Food Techno Co., Ltd.
Tokyo, Japan

Kluwer Academic / Plenum Publishers
New York, Boston, Dordrecht, London, Moscow

ISBN 0-306-48207-X

233 Spring Street, New York, New York 10013

http://www.wkap.nl/

10 9 8 7 6 5 4 3 2 1

A C.I.P. record for this book is available from the Library of Congress

Printed in the United States of America

DEDICATION

The authors gratefully and respectfully dedicate this volume to Dr. Tsuneo Kada.

Tsuneo Kada was born in Tokyo in 1927 and received his undergraduate degree from the University of Tokyo in 1950 and his doctorate in Natural Sciences from University of Paris in 1963 for studies on the mechanism of mutation induction by X-rays and ultraviolet light in *Escherichia coli.* He came back to Japan in 1967 and served as Professor in the National Institute of Genetics, Mishima.

He has promoted the active studies on the genetic mechanisms of mutation induction by radiation and chemicals using bacteria, experimental animals to cultured mammalian cells. He will long be remembered for the numerous scientific contributions (32 books and 288 papers), including the development of the widely used "rec-assay" method to detect the new mutagens easily. He was one of the pioneer in the area of studies on antimutagenesis and anticarcinogenesis in the world.

In 1985 Tsuneo Kada's group involving Hara discovered an inhibitory activity of green tea against spontaneous mutations in *Bacillus subtilis* , and found that the effective substance in green tea was four catechins involving epigallocatechin gallate (EGCg).

He received many prizes including the Japan Agriculture Prize, the Yomiuri Agriculture Prize, etc. He died in November in 1986.

ACKNOWLEDGMENTS

This volume is an English version of "Why is tea beneficial to human health --- the secret of catechin power ---" written by the same authors originally published in Japanese from the Shokabo Publishing Co., Ltd, Tokyo. The authors express their gratitude for giving the permission to publish this English version.

The authors are grateful to many scientists in the Food Research Institute of Mitsui-Norin .Co., Ltd., Fujieda, Shizuoka, for providing many useful research results cited in this volume.

The authors also wish to express their appreciation to Doctor Delbert M. Shankel, Professor of Microbiology and Chancellor Emeritus, University of Kansas, for his continued interest and kind suggestions in the initial manuscript. The authors are indebted to Ms. Andrea Kay Suzuki for her contribution and helpful suggestions in preparing the manuscript. Special thanks are expressed to Miss Yuko Takada, for her technical assistance in preparing the manuscript.

1. VARIOUS KINDS OF GREEN TEA

GYOKURO

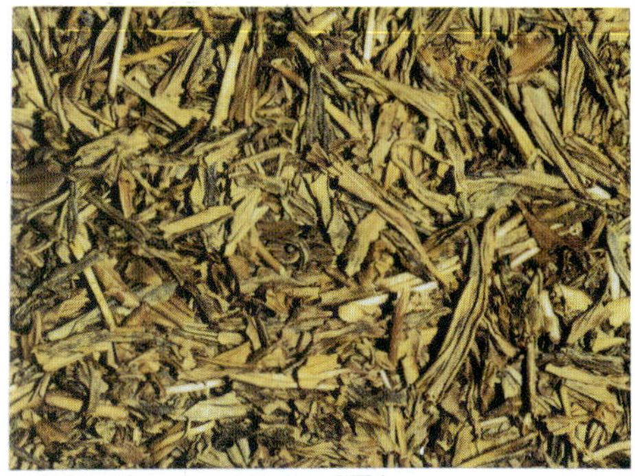

HOUJICHA
(roasted green tea))

SENCHA

GENMAICHA
(popped rice and tea)

BANCHA

MATCHA (TENCHA)
(ceremonial powdered tea)

2. HOME LAND OF GREEN TEA

EXTENDED GREEN TEA FIELD

NEW BUDDING OF GREEN TEA

A GARDEN IN OCHA-NO-SATO
(tea museum)

WATER CULTURE OF GREEN TEA

MANUAL TEA PROCESSING

3. BACTERIOCIDAL EFFECT OF TEA IN THE FORMATION OF DENTAL CARIES

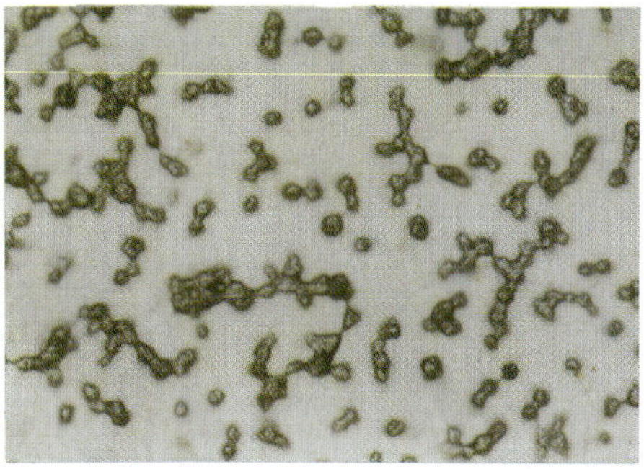

DENTAL CARIES FORMED BY BACTERIA

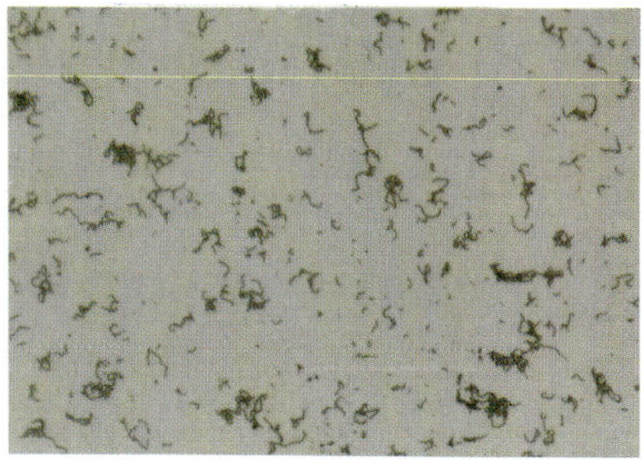

AFTER ADDITION OF BLACK TEA CATECHINS

4. PREVENTION OF DISCOLORATION OF FOODS

DRIED HORSE-MACKEREL FILLETS

NOT SPRAYED SPRAYED BY CATECHIN

SALMON STEAKES

NOT SPRAYIED

SPRAYED BY CATECHIN

PREVENTION OF DISCOLORATION IN ORANGE JUICE
(EXPOSED TO LIGHT FOR 1 MONTH)

After exposure to light for 1 month

5. FOOD AND INDUSTRIAL APPLICATION OF TEA CATECHINS

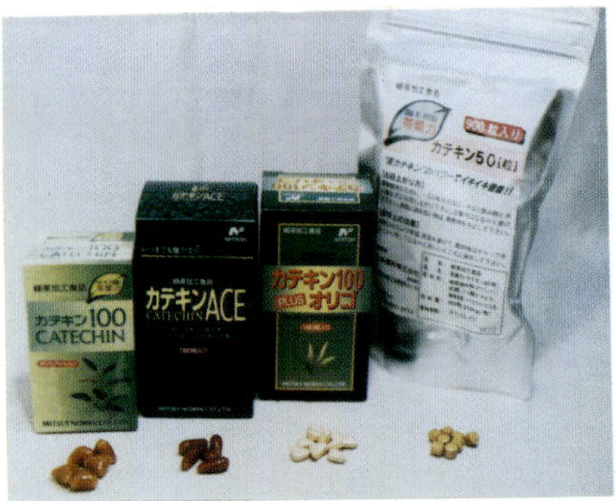

CATECHIN CAPSULE

CATECHIN CANDY

CATECHIN EGGS

CATECHIN FILTER
IN AIR STERILIZER

CONTENTS

1. GREEN TEA IN JAPAN...1
1. 1. WHERE DID GREEN TEA ORIGIINATE?...1
1. 2. BUDDHISM AND GREEN TEA...3
1. 3. GREEN TEA IN FEUDAL TIMES...4
1. 4. ESTABLISHMENT OF THE TEA CEREMONY...5
1. 5. GREEN TEA AND JAPANESE PEOPLE...6
1. 6. PRESENT STATE OF GREEN TEA IN JAPAN...8

APPENDIX 1 GREEN TEA, BLACK TEA AND OOLONG TEA...9

2. PREVENTIVE EFFECTS OF TEA AGAINST HUMAN DISEASES...11
2. 1. TEA IS EFFECTIVE AGAINST MANY DISEASES...11
2. 2. PREVENTION AGAINST CARCINOGENESIS...12
2. 2. 1 The main cause of death in Japan...12
2. 2. 2. The mechanisms of carcinogenesis...13
2. 2. 3. Carcinogenic process in normal cells...15
2. 2. 4. Carcinogens present in environment...18
2. 2. 5. Carcinogens present in food materials...20
2. 2. 6. Anticarcinogens present in food materials...23
2. 2. 7. Tea drinking and stomach cancer...25
2. 2. 8. Tea ceremony instructors and tea...28
2. 2. 9. Antimutagenic activity of tea...28
2. 2. 10. Anticarcinogenic activity of tea...30
2. 2. 11. Programming cell death (Apoptosis) induced by tea...36
2. 2. 12. Inhibition of angiogenesis by tea...36
2. 2. 13. Inhibition of *Helicobacter pylori* growth by tea...37
2. 2. 14. Clinical chemopreventive trials with tea catechins...38

2. 3. PREVENTION AGAINST ALLERGY 38

APPENDIX 2 COLOR, TASTE AND AROMA 39

3. AGE-RELATED DISEASES AND TEA 41
3. 1. PREVENTION OF HYPERTENSION 41
3. 1. 1. Cause of hypertension 41
3. 1. 2. Hypotensive action of tea drinking in animal experiments..42
3. 1. 3. Hypotensive action of tea drinking in humans..44
3. 2. PREVENTION OF DIABETES. 45
3. 2. 1. Cause of diabetes. 45
3. 2. 2. Insulin and sugar-resolving enzymes. 45
3. 2. 3. Tea lowers blood sugar levels in animals 47
3. 2. 4. Tea lowers human blood sugar levels..48
3. 3. PREVENTION OF OBESITY..49
3. 4. SUPPRESSION OF ARTERIOSCLEROSIS..51
3. 4. 1. Mechanisms of arteriosclerosis..51
3. 4. 2. What is cholesterol?..51
3. 4. 3. Good cholesterol and bad cholesterol in the blood..51
3. 4. 4. Tea reduces bad cholesterol 52
3. 4. 5. Effect of tea on human blood cholesterol..55
3. 5. PREVENTION OF SENILE DEMENTIA 56
3. 5. 1. Tea increases mental activity 56
3. 5. 2. Alzheimer's disease... 56
3. 5. 3. Tea prevents Alzheimer's disease..57

APPENDIX 3 STORAGE OF TEA 59

4. INFECTIOUS DISEASES AND TEA 61
4. 1. PREVENTION OF INFLUENZA INFECTION 61
4. 1. 1. Preventive action of tea against influenza virus infection 61
4. 1. 2. Prevention of influenza virus in mice and pigs 62
4. 1. 3. Tea prevents human influenza 62
4. 2. PREVENTION OF DENTAL CARIES AND BAD BREATH 63
4. 2. 1 Prevention of dental caries in school children 63
4. 2. 2. Inhibition by tea of the formation of dental caries 65
4. 3. PREVENTION OF FOOD POISONING 67
4. 3. 1. Foodborne pathogenic bacteria. 67
4. 3. 2. Bacteriocidal activity of tea..67
4. 3. 3. Tea kills bad bacteria and promotes good bacteria 69
4. 4. IMPROVEMENT OF INTRAINTESTINAL BACTERIA AND PREVENTION OF ODOR 70
4. 4. 1. Effect of tea catechins on intraintestinal bacteria..71
4. 5. . PREVENTION OF INTRA-HOSPITAL INFECTIONS 72
4. 5. 1. Intra-hospital infective bacteria, MRSA 72
4. 5. 2. Tea kills MRSA 73

4. 6. PREVENTION OF ALLERGY 73
4. 6. 1. Mechanisms of allergic reactions 73
4. 6. 2. Oolong tea prevents allergic reactions 75
4. 7. PREVENTION OF INFECTION WITH HIV 75
4. 7. 1. HIV 75
4. 7. 2. Tea inhibits the growth of HIV 75
4. 8. PREVENTION OF FORMATION OF WARTS AND INFLAMMATION 76
4. 8. 1. Warts and inflammation 76
4. 8. 2. Condyloma in human genital organs 77
4. 8. 3. Prevention of human condyloma 77

APPENDIX 4 PROVERBS ABOUT TEA 79

5. ANTIBACTERIAL AND ANTIOXIDATIVE ACTIVITY OF TEA 81
5. 1. ANTIOXIDATIVE ACTION OF TEA....81
5. 1. 1. Active oxygen in the body 81
5. 1. 2. Antioxidant action of tea 82
5. 2. PREVENTION OF OXIDATION IN PROCESSED FOODS....82
5. 2. 1. Anti-oxidation of foods 82
5. 2. 2. Preservation of color in foods 83
5. 2. 3. Deodorant action of unsavory meat flavors 84
5. 3. PREVENTION OF OFFENSIVE ODORS BY TEA 85
5. 3. 1. Prevention of odors in the home 85
5. 4. AIR CLEANING ACTION OF TEA 85

6. HOW TO MAKE A GOOD CUP OF GREEN TEA 89
6. 1. SUITABLE WATER FOR GREEN TEA 89
6. 2. CHOOSING GREEN TEA 90
6. 3. HOW TO POUR A CUP OF GREEN TEA 90
6. 4. KINDS OF GREEN TEA 90
Gyokuro, Sencha, Bancha, Houjicha, Kukicha, Konacha
6. 5. TABOOS IN TEA DRINKING 91

7. FOOD AND INDUSTRIAL APPLICATIONS OF TEA CATECHINS 93
7. 1. CATECHIN CAPSULES....94
7. 2. CATECHIN EGGS 95
7. 3. CATECHIN CANDY 95
7. 4. CATECHIN TABLETS FOR BAD BREATH 95
7. 5. CATECHIN DRINKS WITH RADICAL SCAVENGING POWER....95
7. 6. CATECHIN FILTER 95
7. 7. CATECHIN COSMETICS 95
7. 8. CATECHIN MEDICAL PRODUCTS 96
7. 9. PRODUCTS FOR MAINTAINING FRESHNESS OF FISHERY PRODUCTS 96
7. 10. DEODORANT SPRAYS FOR KITCHENS 96

7. 11. CATECHIN-CONTAINING ZINC TABLETS 96
7. 12. SOFT DRINKS CONTAINING CATECHINS 96
7. 13. CATECHIN-CONTAINING INSTANT GREEN TEA 97
7. 14. CATECHIN TEA BAGS 97
7. 15. SLUG REPELLANT SPRAY 97
7. 16. CLOTHES WITH ANTIBACTERIAL AND ANTIOXIDANT PROPERTIES 97
7. 17. FORMALDEHYDE ABSORPTION BY CATECHINS 97

CONCLUDING REMARKS 99
REFERENCES 103
AUTHOR INDEX 107
SUBJECT INDEX 109

FOREWORD

During the course of the last two decades, the health effects of tea and its catechins have been documented in numerous scientific studies and the scientific basis of these effects has been elaborated.

Professor Kuroda and Dr. Hara provide in this volume a translation of a thorough and extensive book published earlier (in Japanese), which will be of considerable interest to scientists working in the field, and to members of the general public interested in natural ways of maintaining health. Both authors are themselves eminent researchers with extensive experience. Dr. Kuroda is best known for his publications employing tissue cultures. Dr. Hara has published numerous research papers on the health effects of green tea and its catechins.

This book provides a comprehensive overview of the historical use of green tea in Japan and elsewhere, followed by a description of the many disease conditions against which these teas have preventive (protective) effects. Of particular interest to "senior citizens" is the chapter on the ability of tea components to ameliorate a wide range of age-related diseases including cancer, arteriosclerosis, and various forms of senile dementias. For scientists and others interested in infectious diseases caused by bacteria and viruses, a chapter dealing with the ability of tea components to inhibit these organisms will be of special interest.

The mechanisms of action of tea and its catechins are not ignored, and there is an excellent discussion of the antioxidative and antimutagenic mechanisms that are significant aspects of the many reactions in which tea components are involved.

The book concludes with discussions of some practical aspects: such as how to make a good cup of green tea. Current and potential industrial applications of the catechins to food products, cosmetics, and other commercial products are also described.

This translation provides both a valid reference book for scientists interested in the field and an informative and easy to read compilation of useful information for the general reader. It should be of broad interest to all who are interested in maintaining healthy life styles.

Delbert M. Shankel
Professor of Microbiology and
Chancellor Emeritus
University

PREFACE

Tea is one of three major beverages which are commonly consumed worldwide; the other two being coffee and cocoa. Types of tea include black tea, oolong tea and green tea. Black tea is a completely fermented product and oolong tea is a half-fermented product of green tea. In Japan, most people drink tea at least three times a day at breakfast, lunch and dinnertime. Tea has become so much a part of everyday life that "tea time" has come to mean the short break that busy people take during their working day.

Most restaurants in hotels and streets in Japan, even those tucked away in narrow streets of the cities, serve tea freely and without charge when customers order their meals. Some have a self-service system where they put out teacups and a teapot on the table along with a thermos of hot water so the customers can help themselves.

Young people frequently use automatic vending machines and convenience stores, which they frequent. Various kinds of bottled or canned drinks are available, along with the usual range of beverages such as juice, Coca-Cola, and other carbonated drinks. In restaurants throughout Japan, at the finish of a traditional meal of sushi, a big mug of tea known as "oagari" is served. On hot sunny days, cracked ice and ice cream containing green tea are refreshing ways to cool down.

During the Edo period, tea became an indispensable drink for the Japanese people. Tea, which holds an intimate place in the daily lives of Japanese people, has recently been found to have surprising beneficial effects on human health, especially on diseases which tend to afflict people as they get older, such as cancer, hypertension, and diabetes.

Originating in South China, tea was imported to Japan, and due to its consumption by Bhuddhist priests, tea developed links with Bhuddism and the feudal society, thus stimulating the spread of its popularity throughout the country. Tea has deeply permeated Japanese culture, giving birth to the tea ceremony and various tea implements.

Recently, it has come to be known that many of the plant foods, which have traditionally been used in Japan, contain components which have marked beneficial effects on humans. Polyphenols in other plants however, may be chemically different from those in tea. Many plant polyphenols have been investigated, but at this stage health effects of polyphenols contained in drinks other than tea can only be surmised based on those found in tea.

Tea is an easily available, low cost beverage, making it one of the most extensively useful drinks there is. Drinking tea not only with meals three times a day, but also at other times, is recommended as being conducive to good health.

This book deals with the preventive effects of tea against diseases such as cancer, hypertension, diabetes, and bacterial and viral infections; and also its antioxidative actions and its ability to promote human health.

The authors hope that this book will be used as a reference for daily life by a great many people in various countries in the world.

1

GREEN TEA IN JAPAN

1. 1. WHERE DID GREEN TEA ORIGINATE ?

Although there is a theory that tea trees were originally grown in Japan, the green tea used at present by Japanese most probably comes from China. Lu Yu, noted Chinese author and tea expert, who was regarded as a "tea god" or "tea ancestor", published a book "Ch'a Ching" (Tea Classic) in A.D. 780. In this book, tea is described as *a good tree in the Sough.* A traditional ancestor of medicinal herbs, "Shen Nung"was referred to as the "Divine Healer" and lived around 2730 BC. He declared that" drinking of tea gives you strength and imparts a relaxing feeling." The place where the tea tree originated may be located in Yunnan in South China.

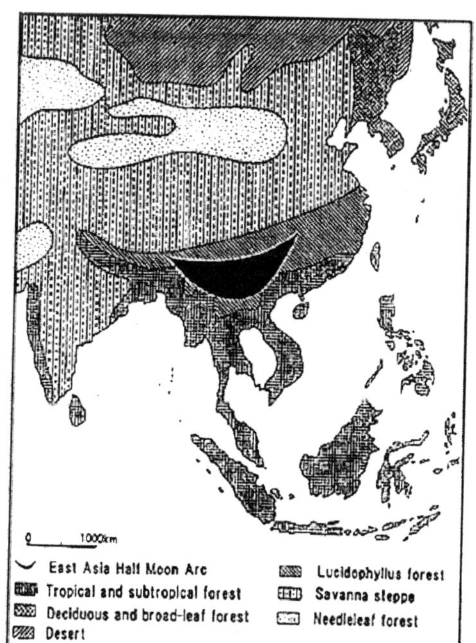

Figure 1. 1. The lucidophyllus forest in Yunnan in China (The East Asia HalfNoon Arc) which is considered to be where the tea tree originated. (Oishi,, S., 1988)

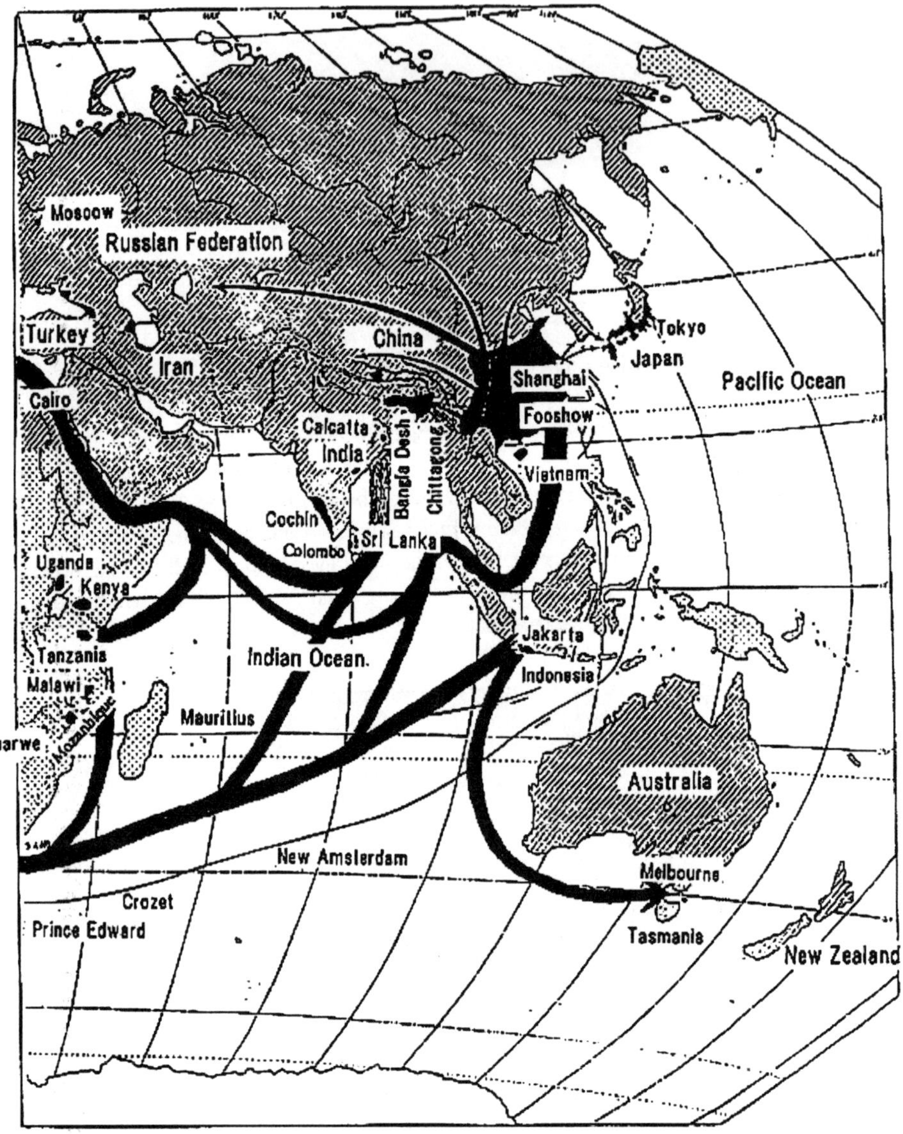

Figure 1. 2. The transport routes of tea from Yunnan in China to cities world wide. (Oishi, S., 1988).

The lucidophyllus forest in the Yunnan district of China is located at the center of the area known as the "half moon arc" (East Asia half moon arc) which stretches from Assam in Western India to Eastern China, as shown in Fig. 1. 1. Tea came to be cultivated throughout Japan from Kyushu, Shikoku and Chugoku (Western area) to the Kinki and Tokai distracts (Central area). Popularity of green tea as a drink spread also to Russia, India and Iran, while in England and Australia black tea became popular (Figure 1. 2)

Tea was introduced to Japan by envoys of the Japanese delegation and several famous priests who had visited Tang and Sung in China. Eichu, Saicho and Kukai, famous priests of high rank, who had introduced Buddhism to Japan, brought tea trees to Japan in the 800s. The first tea garden was planted in the Heian period, according to the order of emperor Saga.

1. 2. BUDDHISM AND GREEN TEA

During the Heian period, the custom of tea drinking gradually diminished, but the ceremony of drinking tea after sutra chanting remained. In the Kamakura period, the connection between tea drinking and Buddhism strengthened again. This popularization of tea drinking was due largely to the activities of a Zen priest, Yeisai, who was a scholar of the Tendai school dogma of the Zen sect located at Mt. Hiei in Kyoto. Yeisai traveled to Sou in China, where he obtained new tea seeds and learned how to make "Matcha" (a kind of powdered tea). After these experiences he returned to Japan, and while carrying out his missionary work for the Zen sect in the Kyushu district, imparted his knowledge of tea drinking and initiated the planting of tea seeds (Figure 1. 3).

Figure 1. 3. Zen Priest, Yeisai, author of the book "Kissa-Yojoki" (Tea and Health Promotion, 1212) (Statue in Makinohara, Shizuoka, Japan)

The name of Yeisai is famous in the history of tea in Japan, particularly because he wrote a book "Kitcha-Yojoki" (Tea and Health Promotion), in which he described the effectiveness of tea in maintaining human health. Yeisai established the Kenjinji temple in Kyoto to pray for the Shogunate Yoritomo Minamoto's death in Kennin, 2nd year (1202). He also made every effort to reconstruct the

Todaiji temple in Nara. He recommended tea as being an excellent medicine to Yoritomo and presented him with his book, "Kitcha-Yojoki, in the 2nd year of Kenpo (1214). "Kitcha-Yojoki" consisted of two volumes, "*Harmony of the five organs*" and "*Escape from devil*" which dealt the with the effects of tea on physical health and mental stability.

"Tea is a hermit drug for recuperation, and extends life expectancy. The mountains and valleys where the tea trees grow are spiritual places. People who consume tea can extend their life expectancy. The people in India and Cathay hold tea in high regard. In Japan too, people like to drink tea. The tea is a *specific hermit drug* lasting from ancient to present time (from "Kitcha-Yojoki").

Yeisai described the effectiveness of tea against various diseases, quoting many Chinese reference books, as follows, in his "Kitcha Yojoki"

1. He claimed that drinking tea is a way of sobering up after indulging in too much sake. He also stated that tea could keep you from sleeping, which is a kind of disease. This notion of sleeping as a sickness may be due to the fact that Yeisai was a Zen priest, and the stimulating effect of tea on the central nervous system is what is thought to be responsible for this action.
2. Tea drinking always produces a feeling of exaltation and the mind is filled with delight and hope.
3. Tea is effective in healing various injuries, promoting urination, and protecting against various diseases.
4. Tea drinking alleviates fatigue.

All these claims may also hold true from the standpoint of present day pharmacological knowledge.

1. 3. GREEN TEA IN FEUDAL TIMES

In the Period of the Northern and Southern Dynasties in Japan, a primitive kind of tea ceremony became fashionable among the samurai warriors who competed with one another in judging the quality of various teas based on taste and aroma. Tea quality evaluation was carried out for a thousand different green teas, initially with ten test drinkings, and sometimes the competition would continue throughout the night with 70 or 100 drinkings going on over 8 or 9 hours. Many court nobles, priests, and warriors took part in these drinking contests and prizes were awarded to winners. After the contests were over, the sake banquets began and many gourmet foods were served. Participants enjoyed drinking sake, singing songs and dancing.

In the 1500s, the Shogun Nobunaga Oda (Figure 1. 4), who does not appear to have had any strong connection with tea in general, enjoyed drinking tea on the battlefield. Nobunaga held big tea parties with members of influential families, warriors, priests and court nobles in attendance, using deluxe tea ceremony implements and tea bowls and serving cakes from China.

In the era of the Shogun Hideyoshi Toyotomi, and as a reflection of this peaceful age, the tea ceremony became popular in Japan and its place in society was established. In the 15th year of Tensho (1587), Hideyoshi organized an unprecedented large-scale tea banquet in the Shrine Kitano Tenjin, which was attended by many tradesmen and farmers.

From the end of the Kamakura age to the Northern and Southern Dynasties, tea trees were planted in many districts including the Kyushu, Kinki and Kanto areas in Japan. Kita-Kyushu in Kyushu, Suruga in Shizuoka Prefecture and Kawagoe in Saitama Prefecture produced great quantities of tea leaves. In terms of the quality of the tea leaves, Togano in Kyoto Prefecture was considered the best, followed by Uji and Daigo in Kyoto Prefecture. Towards the end of the Kamakura age, the purpose of drinking tea was divided between the "Kitcha-Yojoki" School described by Yeisai and tea drinking as a hobby and a means of social intercourse for court nobles and warriors.

Figure 1. 4. The Shogun Nobunaga Oda, who held a large scale tea party in Azuchi Castle in Kyoto (1576).

1. 4. ESTABLISHMENT OF THE TEA CEREMONY

The tea ceremony was established by Noami, a cultured man who lived in the Muromachi age. He reformed the tea ceremony, that was being performed by warriors at that time, by incorporating elements of beauty, form, and outward solemnity. He recommended his apprentice, a priest named Shuko, to the Shogun Yoshimitsu Ashikaga as a teacher of the tea ceremony. At a later time, during the Civil War era, Rikyu (Fig. 1. 5), who was a head of the tea ceremony in the service of the Shoguns Nobunaga Oda and Hideyoshi Toyotomi, nominated Shuko as the founder and the pioneer of the tea ceremony.

Sen-no Rikyu completed the theoretical side of the tea ceremony through the establishment of a new system, and thus, from the room and the utensils used for the ceremony to the manner in which the tea was served; he raised the tea ceremony to a spiritual, cultural level. The ardent activities of Sen-no Rikyu earned him respect as the founder and the guardian deity of the tea ceremony. Three schools of the tea ceremony; Omote-senke, Ura-senke and Musha-koji emerged from this period, and these schools have continued to the present time as the orthodox schools of the tea ceremony in Japan.

The tea ceremony came to be entwined with the doctrine of the Zen sect and the religious aura of its structure was epitomized in its austerity. In his reform of the tea room, Sen-no-Rikyu replaced the extravagance of the tea ceremonies of the previous era with an austerity that was exemplified in the bareness of the four and a half-mat room. The construction of the room was Spartan, the walls made of unfinished plaster, the wood of the pillar rough-cut, and the windows, irregular in shape, were no more than a hole in the wall. He also introduced the "nijiriguchi", making the entrance to the room so low that the guests, regardless of their social status, had to stoop down and crawl into the tearoom; thus transcending the very distinct class divisions of the feudal age. In keeping with the simplicity of the surroundings, the meal served was unassuming, one soup and one vegetable dish, while the seasonal flowers arranged in the alcove were in no way imposing. The garden was known by a special name "roji" or "dewy ground", through which the guests attending the tea ceremony would pass before they entered the tearoom. In the act of walking through the garden with its stepping stones, hedge, wicket gate, and washing their hands and rinsing their mouths at the stone basin, the guests were undergoing a purification ritual with the purpose of throwing off the mundane and attaining spiritual peace.

Figure 1. 5. Sen-no-Rikyu, who established the system of the tea ceremony (1522-1591).

The tea ceremony is not the only basis for the making and drinking of tea. Tea is also heart warming when served in solitude and quiet.

During the Edo age in Japan, there were four distinct social strata: military, agricultural, industrial and merchant classes. The military class, which held the highest position in the feudal system, developed their own tea drinking rituals. Enshu Kobori, a teacher of the tea ceremony during the time of the Third Shogun Iemitsu Tokugawa, was the head of a castle in the area producing 3,000 koku (450 tons) of rice.

He designed a tea room with wide windows making it brighter, and established standard procedures for the tea ceremony, such as the formal etiquette for making tea, the arrangement of tea utensils, hanging of scrolls, and dress codes, all according to the social standing of the Shogun, the court nobles, and the nuns of convents of the Imperial family. In this feudal social class system, under the administration of Shogun Tokugawa in the Edo age, the system, of having a head of each school of the tea ceremony was established.

1. 5. GREEN TEA AND THE JAPANESE PEOPLE

Distinct from the tea ceremony developed under Shogun and Daimyo, rulers of the districts in the feudal period of the Edo age, tea drinking gradually became part of the lives of the common people. In the Nambu Clan (Aomori Prefecture in the north part of Japan), the tea seeds obtained from Yamato

(Nara Prefecture) were planted in Keicho 13th year (1608). The tea seed planting extended gradually to the Hikone Clan (Shiga Prefecture) and Hiuga-Miyako-no-jo Clan (Miyazaki Prefecture). A tea garden was established by Ryukyu Clan (Okinawa Prefecture) in 1673, and the cultivation of tea trees spread throughout Japan, from the northern parts to the south.

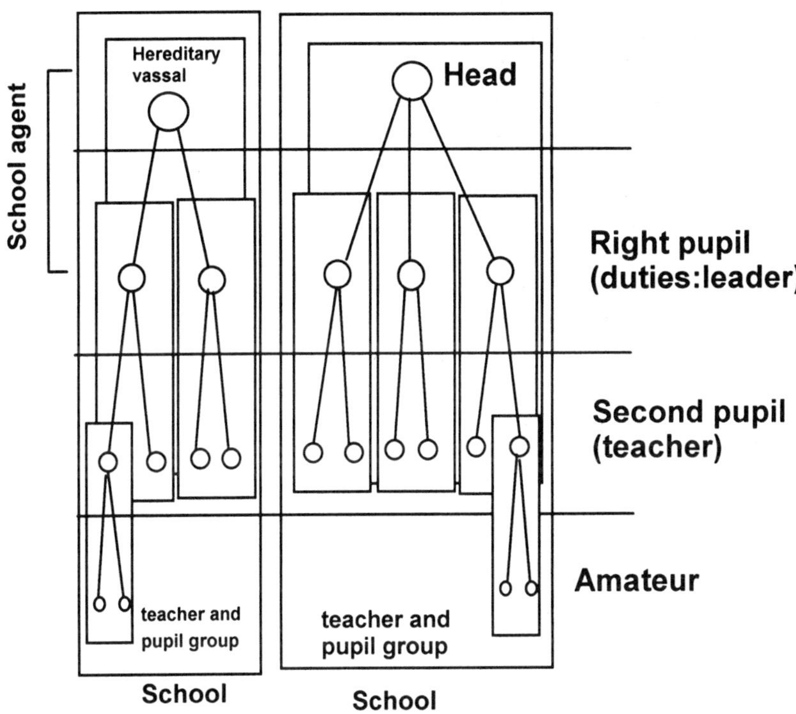

Figure 1. 6. The master system of the tea ceremony.

Since teachers of the tea ceremony were from the ordinary class, economic power moved from the military class to the common people. The tea ceremony of the common people, which was not based on the military class but instead was based on the wider economic standing of the people, was developed. Thus, the typical family school systems for the tea ceremony in the so-called four Kyoto Schools, such as the Sanzen School in Kamigyo (Kyoto Prefecture) and the Yabuuchi School in Simogyo (Kyoto Prefecture), were established (Figure 1. 6)

The issuance of a diploma, a certificate for the tea ceremony teachers in these family school systems, provides economic and ideological support. These diplomas are generally classified into 4-16 classes such as Shoden (first transfer class), Chuden (middle transfer class), Okuden (high transfer class), Hiden (secrete transfer class) and Menkyo-Kaiden (final class), requiring payment of permission fees.

The permission fees are paid as an income to the head of the school family and the disciples. The relationship between the head masters and their disciples appears to be rather like the systematic structure between the lord and his retainer, or the retainer and the next retainer, in the military class in Japan. The diplomas, which have almost an aura of mistique, are exclusively issued by the hereditary head master.

1. 6. PRESENT STATE OF GREEN TEA IN JAPAN

After a long time of national isolation during the Edo period, the export of Japanese tea began in the Meiji period, firstly to England and the United States, and then to other countries worldwide. Soon after this, though, the export of Japanese tea all but ceased. This was probably due to the lack of knowledge on the part of Japan at that time, on what kinds of tea and to which foreign countries they should export. There was also insufficient knowledge of the practical aspects of trading such as setting the price, insurance, foreign exchange orders for sending goods and the procedures for receiving payment.

But in spite of this inexperience, Japanese tea was the second major export product after raw silk. More than 80% of exported tea was directed to the United States and about 10% was to Canada. The majority of this tea was green tea. In 1877 green tea was Japan's star export product, occupying 22% of the total exports, and was the main product to bring in foreign currency.

In 1882 India started to export black tea to the United States. This was the beginning of fierce competition between the black tea from India and the green tea exported from Japan and China. In 1893 the International Exposition was held in Chicago, and in the wake of this event a large amount of black tea flowed from India and Ceylon (Sri Lanka) to the United States. In consequence the green tea exported from China was greatly affected, but during this time the green tea exported from Japan increased slightly in amount.

The preference in America gradually changed from green tea to black tea, but the greatest damage to Japanese tea exports was due to the Second World War, which began in 1945. The amount of Japanese tea for export was restricted. Food imports into Japan stopped completely, and so, due to the lack of food, the land used to produce tea was converted into the production of food products. Thus, the tea industry in Japan, which had grown since the Meiji period, was totally destroyed.

After World War II, the situation of food supplies in Japan recovered gradually and tea began to play an important role as a collateral product for imported foods. In an attempt to make the tea fields, which had been reduced in size and were poorly managed, more productive, research stations and institutions were established throughout Japan. Today Japan can be proud of a modernized tea industry, with the popularization of the excellent tea variety "Yabukita", which is now the variety produced in the largest amount in Japan. Morecover improvements in production have been made in the areas of fertilization, the use of agricultural chemicals to protect against infectious and harmful insects, management of tea fields, mechanization of the process of harvesting the leaves, and the improvement and enlargement of the machines used for processing.

APPENDIX 1

GREEN TEA, BLACK TEA AND OOLONG TEA

Tea is generally divided into three groups; green tea, black tea and oolong tea. All of these groups of tea are produced from "tea" (*Camellia sinensis*) leaves. Green tea is produced by completely inactivating the enzyme, polyphenol oxidase, in tea leaves by exposure to steam vapor for 30 seconds, after which it is rolled (crushed) by hand. Green tea is thus known as unfermented tea since this inactivation of the enzyme ensures that fermentation does not take place. Black tea is produced by treating fresh tea leaves at temperature of 28°C for 18 hours, thus reducing the water content by half. It is then rolled vigorously by machine and fermented at 25 to 30°C for 2 - 3 hours. During this time the tea leaves are still living and it is because black tea is produced according to these processes that it is called "fermented tea". However, this fermentation process is not due to microorganisms, but is brought about by the oxidation of catechins by oxidative enzymes present in tea leaves. During the process of fermentation, the color of tea changes to a reddish tone. Oolong tea is produced by a process in which fresh tea leaves are half fermented by the enzyme present in tea leaves. Thus oolong tea is called "half fermented tea".

RYOKUCHA (GREEN TEA)

The terms "Japanese tea" and "green tea" are synonymous since this is the tea which is consumed so abundantly in Japan. Depending on the conditions during cultivation, as well as differences in the processing procedures of the tea leaves after picking, many types of green tea are produced. The optimum temperature for brewing tea, as well as its taste and fragrance are different depending on the kinds of tea.

SENCHA

Sencha is the type of tea most widely drunk by Japanese people. Young buds of tea leaves, grown with exposure to plenty of sunshine, are picked, steamed and processed completely. Sencha is a refreshing tea with a moderate mix of sweet and astringent tastes.

BANCHA

Bancha is produced from leaves remaining after the young buds of tea leaves have been picked and also from leaves grown in late Autumn which are somewhat hard. Hard leaves and stalks left over from production of Gyokuro and Sencha are also materials for Bancha. Bancha made from stalks only is called Kukicha (stalk tea) or Boucha (stick tea).

OOLONG TEA

Oolong tea is described in Chinese as "鳥龍茶". Oolong tea is tea, which has been half fermented by an enzyme present in tea leaves. Oolong tea has the same content of caffeine and polyphenols as green tea, but has no vitamin C content. This tea is produced in Fukien, Kwangtung in China and in Taiwan. There are many different varieties of oolong tea which are produced according to varying production procedures. All oolong teas sold in Japan are imported from outside countries.

BLACK TEA

Black tea is the product of the fermentation of polyphenols by the oxidizing enzyme present in green tea leaves. The origin of black tea is China. The Assam strain was found in Assam in Bangladesh and cultivated in India, Sri Lanka and other countries. Japan does not produce black tea on a commercial basis.

2

PREVENTIVE EFFECTS OF TEA AGAINST HUMAN DISEASES

2. 1. TEA IS EFFECTIVE AGAINST MANY DISEASES

Tea, which was discovered in China in BC 3000 or earlier, has been regarded as a beverage with medicinal effects, which imparts a good feeling to both mind and body. Tea was introduced to Japan from China, and with the establishment of tea fields throughout the country the popularity of tea among Japanese people grew.

In his book "Kissa-Yojoki" written in the Kamakura Age, Yeisai lauded the physiological and medicinal effects of tea, such as its stimulating effect on the central nervous system, revival from fatigue, improvement of bowel movements and diuretic effect. However, for the Japanese people today, tea is not a special medicine, but a beverage which is consumed regularly in their daily lives.

Medicinal and pharmacological effects of tea have thus been known from very early times, while in relatively recent years more detailed research on the compounds in tea has been carried out. Caffeine was found in tea in 1872 and the presence of tannin was confirmed in 1948. Various studies on the physiological functions of tea were carried out in the 1980s, and over these last twenty years a surprising number of medicinal and physiological effects have been revealed.

Unlike coffee and cocoa, tea is astringent, the compounds responsible for this astringency being catechins, which are a kind of polyphenol. Other chemical components such as theanine and vitamin C were successively identified and their medicinal effects were scientifically clarified. Among these tea components, polyphenols in particular have many medicinal and phamacological effects on various conditions which are likely to occur as people get older. As shown in Table 2. 1, these include cancer, high blood pressure, diabetes, obesity and increased cholesterol levels.

Furthermore, tea can protect against infections by viruses such as the influenza virus, and HIV (human immune deficiency virus); and also against the bacteria which cause tooth decay. The compounds such as caffeine, vitamins C and E, and carotene, which are contained in tea are effective in protecting against various ailments and thus maintaining health. Studies on tea have been carried out in universities and laboratories in Japan and many countries throughout the world.

Table 2.1. Components in green tea and their effects on human health and diseases (Kumakura, I., *et al.*, eds. 1999)

Components	Contents	Functions and Effects
Water-soluble		
Catechins	10 - 18%	Antioxidativity, antimutagenicity, anticarcinogenicity, suppression of high blood pressure, suppression of high blood sugar, suppression of platelet agglutination, antibacterial, anti-dental caries, antiviral, improvement of intestinal bacterial flora, anti-allergy, deodorization
Caffeine	2 - 4%	Central nerve stimulation, prevention of sleepiness, cardiotonic effect, diuretic effect, promotion of metabolism
Theannin	0.6 - 2%	Lowering of blood pressure, regulation of brain and nerve functions
Flavonols	0.6 - 0.7%	Increase of capillary blood vessel resistance, antioxidativity, anticarcinogeneicity, prevention of heart diseases, deodorization
Complex polysaccharide	ca. 0.6%	Suppression of high blood sugar
Vitamin C	0.15 - 0.25%	Antiscorbutus, antioxidativity, anticarcinogenicity, prevention of influenza infection, prevention of cataract, strengthening of immune system
γ-Aminolactic acid	0.1 - 0.2%	Lowering of blood pressure, regulation of brain and nerve functions
Saponin	0.1%	Anti-asthma, antibacterial, lowering of blood pressure
Vitamin B_2	12 ppm	Prevention of angular stomatitis and dermatitis
Minerals	3 - 4%	Zn, Mn, Cu and Se : antioxidant, Zn: prevention of taste abnormality, (inability to distinguish different tastes), prevention of immune activity fall, prevention of dermatitis F: prevention of dental caries
Aroma components	10 - 20 ppm	Aromatherapy effect
Water-insoluble		
Food fibers	ca. 30%	Prevention of life-style related diseases such as cancer, heart diseases, etc.
Proteins	ca. 24%	Nutrient
Crude lipids	3.4 - 4%	Nutrient
Chlorophyll	0.6 - 1%	Anticarcinogenicity
Vitamin E	0.02 - 0.07%	Prevention of hemolysis, suppression of lipid peroxidation, anti-carcinogenicity, antidiabetes, prevention of cataracts, potentiation of immune system
β-Carotene	ca. 0.02%	Antioxidativity, anticarcinogenicity, potentiation of immune system

1 ppm=10^{-6}

2. 2. PREVENTION AGAINST CARCINOGENESIS

2. 2. 1. The main cause of death in Japan

Before World War II, the average human life span was said to be fifty. Now though, the average life span of Japanese has reached more than 83 years old for females, and more than 77 years old for males, making Japan the country with the longest average life span in the world. In Japan many people live to be more than 100 years old. Twin sisters "Kin-san" and "Gin-san" (gold and silver in Japanese) who lived to be well over 100 became celebrities in Japan because they were so active and healthy. In the field of entertainment arts, many dancers, novelists, are painters are continuing to work actively even over the age of 90 years. Future generations may do even better.

Recently, in Japan, the birth rate has been dropping while the percentage of the population over the age of sixty has been increasing; and to cope with this situation measures need to be taken in social, economical and medical fields.

This rapid increase in the average life span may be due to a number of factors such as developments and improvements in medical procedures, improved hygiene, a dramatic fall in the number of fatalities in young people infected with tuberculosis due to the use of antibiotics such as penicillin and streptomycin, and a decrease in the infant death rate. As shown in Figure 2. 1, the number of deaths caused by tuberculousis decreased rapidly after 1945, while the number of deaths caused by cerebral apoplexy due to high blood pressure increased.

Since 1950, the second cause of death has been cancer (malignant neoplasia), and the third was heart disease. Deaths caused by cerebral apoplexy has now decreased due to developments in prevention and treatment of this disease, and since 1985 the number one cause of death has been cancer. In 1995, 28% of Japanese died from cancer. One of the reasons for the increased cancer death rate may be the larger proportion of older people in the population. Techniques for detecting cancer at earlier stages and treatment procedures have improved considerably, but in the field of prevention of cancer there has not been so much advancement.

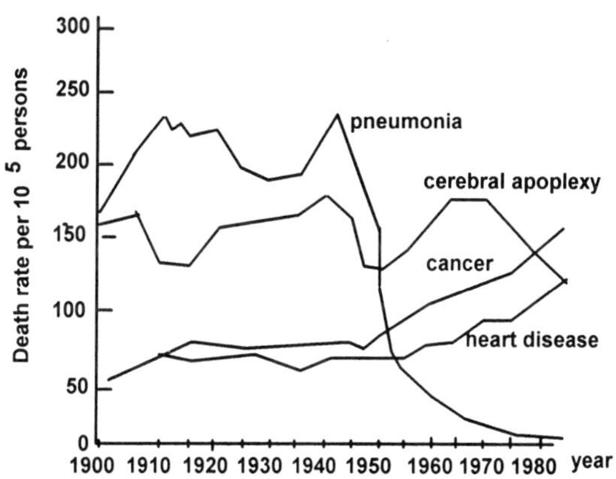

Figure 2. 1. Changes in the cause of death of Japanese in the 20'th century

2. 2. 2. The mechanisms of carcinogenesis

What is the cause of cancer? Is it hereditary factors that are passed down successively over the generations? Or, is it rather a result of exposure to certain carcinogens that are present in the environment? Based on the data obtained from many studies, it has been determined that 20% of cancers in men and 25% of cancers in women may be caused by hereditary factors or genetic make-up inherited from parents (Figure 2. 2). However, the majority of cases of cancer may be caused by any number of carcinogenic substances contained in, for example, the atmosphere, water, soil, foods, drinks, food additives, medicines, etc. as well as X-radiation and UV radiation.

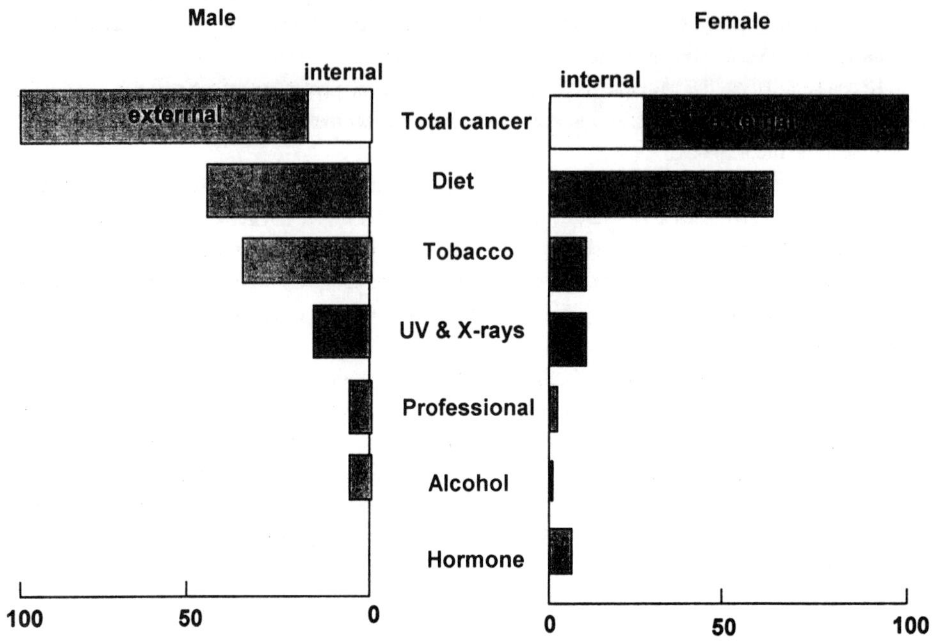

Figure 2. 2. Various factors as possible cause of cancer. (Wynder , E. L..and Gori,,E. G. B., 1977)

What percentage of carcinogenesis is caused by these respective environmental factors? The results of investigations on the causes of cancer in American people, which were carried out by the Royal Cancer Research Foundation in England, are shown in Figure 2. 3. Food and tobacco were responsible for the greatest number of cancer cases, causing respectively one third of all cancers. The causes of the remaining one third of cases included infectious bacteria and viruses, sexual (behavior), professional hazards of dealing with carcinogens, alcohol, industrial products, medicines, food additives, X-radiation, UV, etc.

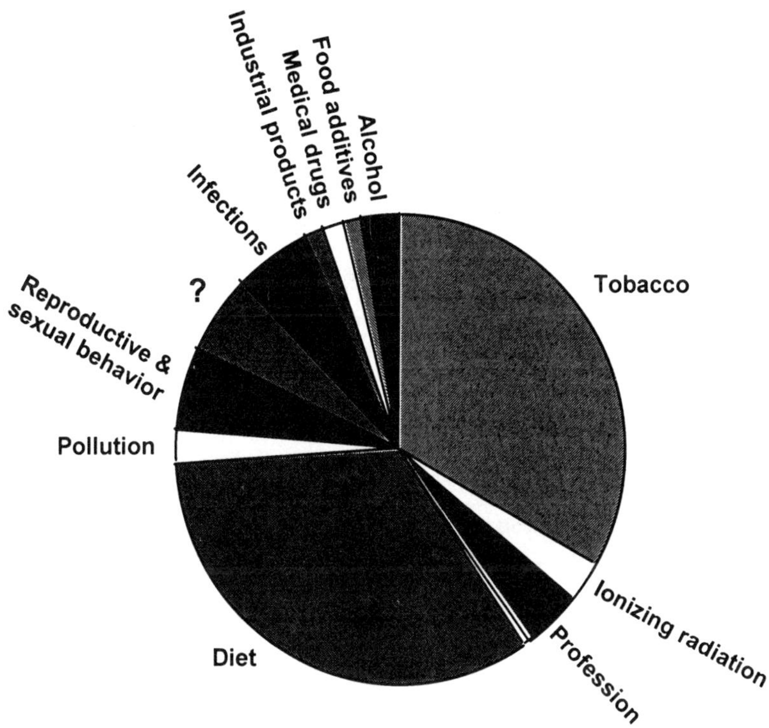

Figure 2. 3. Environmental factors as possible causes of cancer. (Kee,, M., 1983)

While alcohol is involved in 3% of all cancer cases, it is not, however, a direct cause of cancer, but rather indirectly increases a rise in the risk.

2. 2. 3. Carcinogenic processes in normal cells

The human body is composed of as many as 6 x 10^{13} cells. These cells form various organs and tissues such as brain, liver, lung, kidney, heart and blood cells, which perform the normal physiological functions in the human body. It only takes one normal cell to change to a malignant cancerous one, which then proliferates rapidly, thus depriving the normal cells of the necessary nutrients and impairing normal functions of the body, until finally death results.

Why do the normal cells change to malignant cancerous cells? Every normal cell has a nucleus in its center. In the nucleus are the chromosomes containing the DNA (deoxyribonucleic acid) that makes up all of the genetic information (genes). Carcinogenic substances and exposure to radiation induce chemical changes in DNA molecules (Figure 2. 4)

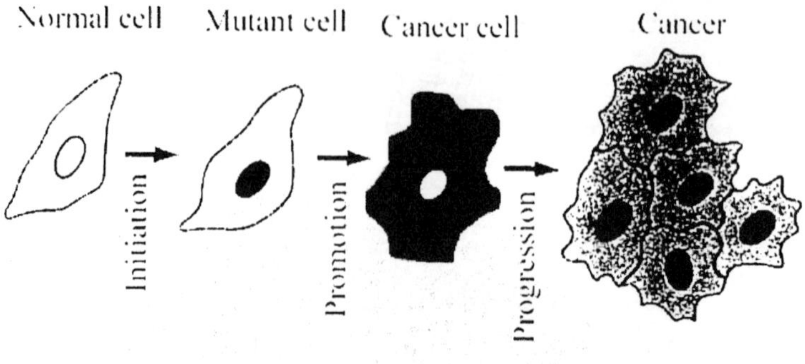

Figure 2.4. The multi-step process of carcinogenesis

Tabel 2.2. Main oncogenes and antioncogenes (Oncogenes and Anti-oncogenes, Toyoshima, K. 1991)

Gene	Function
Oncogenes	
sis	Platelet-derived growth factor (PDGF)
hst-1	Heparin-binding growth factor (HBGF)
int-2	"
erbB	Epidermal growth factor (EGF) receptor
sam	Fibroblast growth factor (FGF) receptor
mos	Meiosis control
ras	Low molecular G protein
myc	Gene expression regulation
fox, fra	"
mos	Angiotensin receptor
Anti-oncogenes	
RB	Retinoblastoma
p53	Li-Fraumeni syndrome
WT	Wilms syndrome
K-rev	G protein
DCC	Colon cancer

This is the first step in carcinogenesis, and is known as "initiation". However, even though there has been this genetic change in DNA of the cells, this alone does not make them malignant cells. The second step is known as "promotion" and occurs in the cell membrane and cytoplasm of the cells. It is thought that this process may involve the genes controlling the cell membrane and cytoplasm.

In the process of "initiation", a change occurs suddenly in the DNA of the cell nucleus, but of the more than 10^5 genes in each cell in the body, in which cells do changes in DNA occur? When normal cells change to malignant cells, it is not just a single gene that is involved. There may be as many as a hundred genes concerned with carcinogenesis and these are known as "oncogenes". These "oncogenes" are generally present in normal cells, and their function is to regulate the normal cell divisions and functions. When these oncogenes mutate, malignant transformations occur in the cells, leading to carcinogenesis.

The main oncogenes found are shown in Table 2. 2. The oncogenes, which mutate to cause malignancies, may differ according to the kind of cancer. In addition there are "anti-oncogenes" which protect the normal cells from becoming malignant. However, the anti-oncogenes too may suddenly mutate, thus causing the normal cells to change into malignant cells.

The development of colon cancer in human is a typical example of carcinogenesis, with successive mutational changes in oncogenes and anti-oncogenes, as shown in Figure 2. 5. In the process of colon cancer, firstly an oncogene, *FAP*, mutates resulting in the hyperplasia of colon epithelium and a decrease in DNA methylation. This is the first step in colon carcinogenesis.

In the second step, another oncogene, *K-ras* mutates resulting in intermediate adenoma, and then the third oncogene, *DCC* causes a deficiency, which results in advanced adenoma. Further an anti-oncogene, *p53* mutates resulting in the loss of tumor inhibitory activity, and finally malignant cancer develops. These carcinogenetic processes involve mutations or deficiencies in at least four oncogenes and an antioncogene

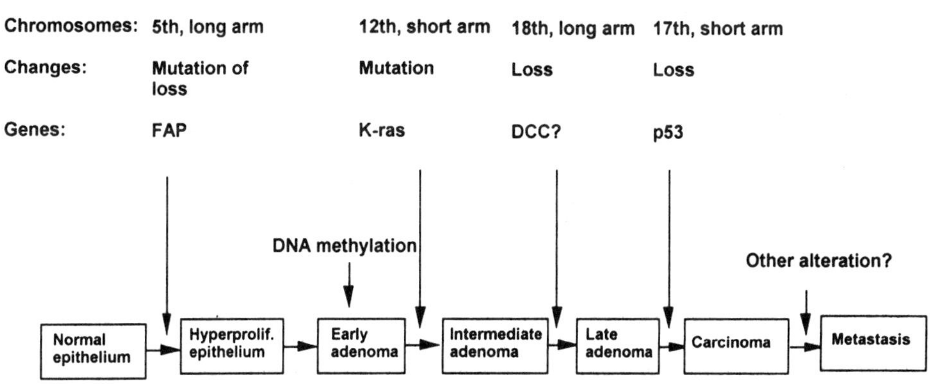

Figure 2. 5. Changes which occur in oncogenes during the process of carcinogenesis of large intestine.(Fearon , M. and Vogelstein, B., 1990)

The anti-oncogene *p53* has recently been used in the gene therapy treatment of cancer. Cancer cells taken from patients have the *p53* gene placed into their nuclei, in an attempt to revert the cancerous cells back to normal ones. Even after the two processes of initiation and promotion have occurred in the cells, there is not a prolific increase in cancerous cells. For the cancerous cells to proliferate, a third step, "progression" is necessary. Although the processes involved in this step are not completely clear, it is thought that substances which accelerate the rejuvenation of blood vessels are excreted from the altered cells, and these blood vessels carry added nutrients thus promoting the proliferation of cancerous cells. The mechanisms of carcinogenesis in normal cells involves these multiple steps, and this is called the "Multistep Theory of Carcinogenesis".

2. 2. 4. Carcinogens present in environment

Chemical substances and radiation, which induce mutations at each step of carcinogenesis, are called "carcinogenic agents". Foods and tobacco, which induce mutations and malignant transformations, are "carcinogenic agents". Among the many carcinogenic agents, substances that induce mutations in DNA are called "mutagens".

Recently many mutagens which we are exposed to in our daily lives, have been detected in foods, drinks, food additives, medicines, cosmetic products, tobacco, exhaust fumes of automobiles, and in the ash and smoke from incinerators.

As already stated, next to food, tobacco is a major cause of cancer. Its carcinogenic activity is due to benzo[a]pyrene and other carcinogens contained in tobacco smoke, and these cause lung cancer and other cancers. The incidence of lung cancer in smokers is 30 to 100 times higher compared to that of non-smokers, as shown in Figure. 2. 6.

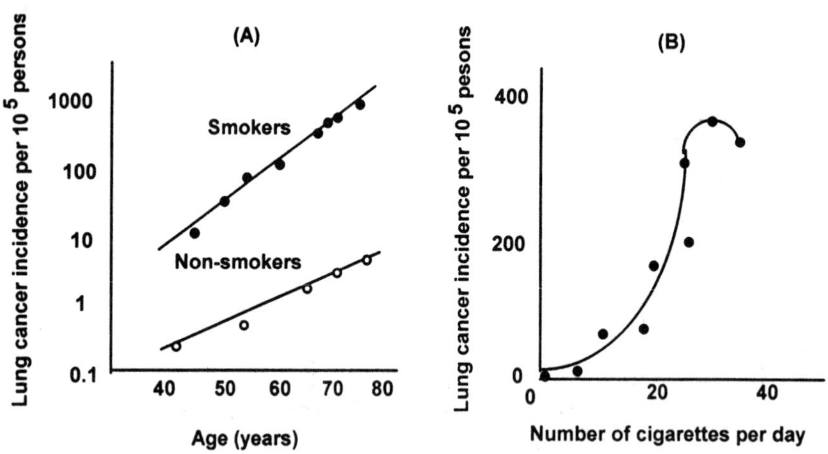

Figure 2. 6. Incidence of lung cancer in tobacco smokers and non-smokers (A) and in comparison to the number of cigarettes a day (B). (Doll, R., 1978).

Moreover, the longer the period of smoking, the greater the risk of contracting lung cancer. The incidence of lung cancer is related to the amount of tobacco smoked daily. The more cigarettes smoked a day, the higher the incidence of lung cancer. Heavy smokers who smoke 35 cigarettes a day, showed about 400 times higher incidence of lung cancer than that of non-smokers. In the United States, the tobacco companies are obligated to print on the outside packaging of their product: "Smoking is harmful to the health and may cause cancer or other diseases". In some cases lung cancer patients have taken their case to court, claiming damages from the tobacco company, and the company has had to pay indemnification.

To detect the carcinogenic activity of substances present in the environment, experimental animals such as rats and mice are used and the test substances are put in their food or applied to the skin in order to examine the carcinogenicity of the substances. To test the carcinogenicity of even just one chemical, many animals are required and the tests are carried out over a period of 2 to 3 years. In order to cut costs and reduce the length of time of testing for carcinogenicity of chemicals, a procedure that tests only the activity of the substances that induce the initiation or promotion steps was developed.

Substances that induce the initiation stage may also induce mutations in DNA, not only in humans, but also in experimental animals such as hamsters, rats and mice, as well as in cultured mammalian cells, plants such as soybeans and *Tradescantia*, insects such as *Drosophila* (fruit fly), *Bombix mori* (silkworm), fungi, yeasts and bacteria such as *Salmonella typhimurium*, *Escherichia coli* and *Bacillis subtilis*. Thus procedures related to short-term tests for carcinogenicity of chemicals using these organisms have been developed, and the carcinogenicity of many environmental chemicals has been able to be tested at a low cost within a few days to a few weeks. These short-term test procedures have enabled the identification of many carinogens and anticarcinogens in the environment which are found in various foods and drinks consumed in our daily lives. Some of the major carcinogens are shown in Table 2. 3.

Table 2.3 Carcinogens found in environment

Environment	Carcinogens	Source materials
Foods	Aflatoxin B	Rice and peanut molds
	Cycasin	Cycard seeds
	Fukinotoxin	Coltsfoot flower
	Trp-P-1	Burnt part of fish and meat
	Trp-P-2	"
	Nitrosamines	2nd class amine (fish and fish eggs) and nitrite (fresh vegetables, pickles)
Atmosphere	Benzo[a]pyrene	Exhaust smoke tobacco
	Dioxin	Burning of vinyl chloride
Life style	Asbestos	Building materials, heat-shielding materials
	Ca	Industrial products
	As	"

After World War II, Japan imported rice from South-East Asia, because of the shortage of rice. Some of this rice was infected with fungi, turning it a yellow color. This rice was called "yellow rice" and was found to contain a carcinogen, aflatoxin B_1. Also, the mutagenicity and carcinogenicity of other substances such as furylfuramide, used as a preservative in tofu, and an edible red dye, phloxine, which was used as a food color was determined and they were withdrawn from the Ministry of Health and Welfare list of permitted food additives.

Fukinotoxin, contained in Fuki-no-tow (the flower of colts food), and amino acid pyrolysis products such as Trp-P-1 (tryptophan pyrolysate), Phe-P-1 (phenylalanine pyrolysate) etc. were also found to be mutagenic according to various test systems. There are also cases where food consumed becomes carcinogenic after undergoing a chemical reaction with other substances in the stomach. For example, amines which are contained in fish, meat and fish eggs is not carcinogenic as a single substance, but it changes to the carcinogenic nitrosamine when it reacts with nitrite contained in fresh vegetables or pickles which are consumed at the same time.

Recently, dioxin which is produced in the process of incineration of vinyl chloride, benzo[a]pyrene, contained in the exhaust fumes of motor vehicles, and smoke from tobacco, have become a great social concern due to their carcinogenicity.

2. 2. 5. Carcinogens present in food materials

In various countries including Japan and the United States, many studies have been carried out which compare the eating habits of people living in certain districts with the rates of stomach cancer. In Japan it was found that areas with a high rate of stomach cancer had a high consumption of rice and salty foods such as bean paste (miso) and pickles. In European countries and the United States, the rate was high when there was a predominance of starch foods, such as potatoes and cereals, and foods with a high salt content such as dried and salted fish (Table 2. 4).

Table 2.4 High risk and low risk foods as a possible causes of stomach cancer

Object	Researcher	High risk foods	Low risk foods
Japanese	Senoki	Staple food of rice only, gluttony	———
"	Hensel	———	Fresh vegetables, fruits
"	Hirayama	Salty foods, fish, pickles	Milk
"	Kamiyama	Rice, salt fish, pickles	Carrots, onions
Hawaiian-Japanese	Hensel	Dried fish, salt fish, pickles	Fresh vegetables, milk
American (New York)	Graham	Potatoes	Fresh vegetables
Norwegian	Bierke	Grain, fish	Vegetables, fruits
Israeli	Modan	Starchy foods	———

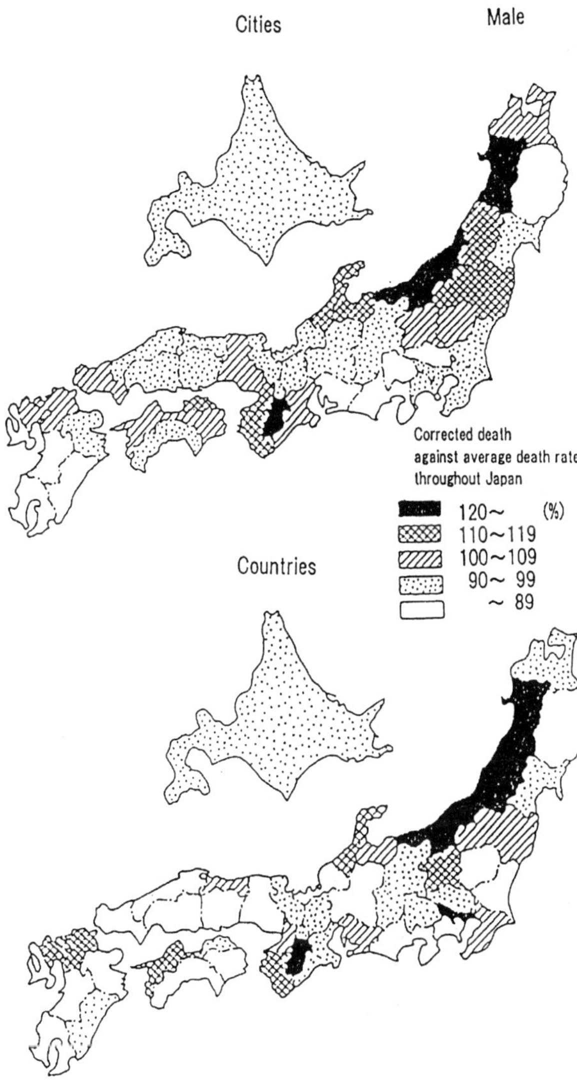

Figure 2. 7. Rate of death from stomach cancer in each Prefecture in Japan: (Average frequency in all Japanese: 100%) (Hirayama, Y., 1978)

On the contrary, in areas with a low rate of stomach cancer, residents were consuming a large amount of fresh vegetables, fruits, and milk. The foods which may contain cancer-causing factors are called "high risk foods for cancer", and those which are not so likely to contain cancer-causing factors are called "low risk foods for cancer".

On comparison of the death rates of stomach cancer patients in each prefecture of Japan, it was found that the areas on the Japanese Sea Coast in the North-East District including Akita, Yamagata and Niigata Prefectures had higher stomach cancer mortality for both males and females as shown in Figure 2. 7. On the other hand, a lower stomach cancer mortality rate was found in some prefectures in Southern Kyushu.

Of interest is Nara Prefecture, where the death rate of stomach cancer patients is high. This may be due to a particular eating habit of the people living in this area. They have a long history of frequently eating a hot rice gruel dish containing edible ferns; and whether it is because of prolonged exposure to heat stimulation or carcinogenic constituents in the plants, these may be factors which contribute to the high rate of stomach cancer.

Shizuoka prefecture is notable for its low rate of stomach cancer mortality. In Shizuoka prefecture the amount of tea production is the highest in Japan. The stomach cancer mortality is about 80% of the average rate in Japan. In particular the area in Shizuoka prefecture, which produces the most tea, has only 20 to 60% of the average stomach cancer mortality of Japan. As already discussed, the kind of foods predominately consumed in the diet greatly influence the incidence of contracting cancer, and further evidence which supports this fact has come to light.

On comparison of cancers in people living in Japan and America, the frequency of stomach and liver cancer in Japanese people was found to be four to six times higher than that of American people living in California (Figure 2.8). On the other hand, the incidence of prostate and colon cancers are one fourth to one fifth lower in Japanese than in Americans. These differences in the site of the cancer may be due to different eating habits between Japanese and Americans: the former having a diet consisting mostly of rice and vegetables and the latter consuming more bread and meat

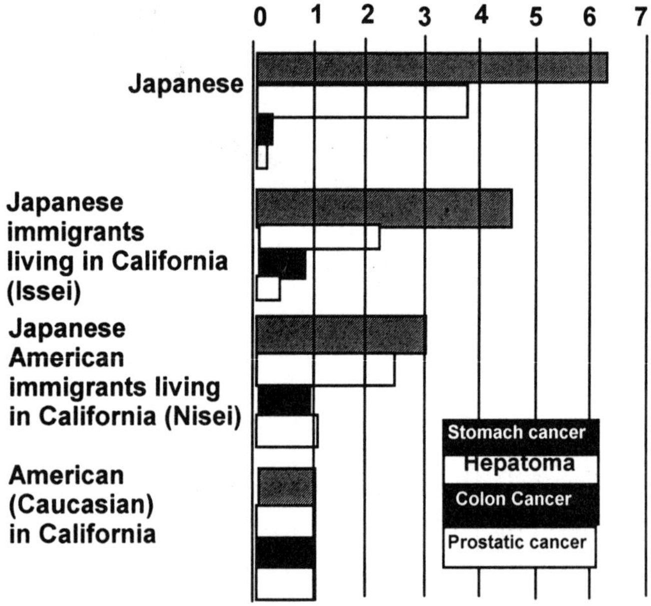

Figure 2. 8. A comparison among Japanese and Japanese-American immigrants, 1 st and 2 n generations of the frequency of cancers in various organs. (Takebe, H., 1991).

Of the most interest, however, is the following fact. After immigrating to California, the first generation of Japanese Americans began to consume an American diet and their incidence of stomach and liver cancer decreased to two thirds and half respectively of that of Japanese living in Japan, while their incidence of prostate and colon cancer approached the level of Americans. Further, in the case of second generation Japanese Americans who were born and grew up in the United States, their eating habits hardly differed from those of native Americans, and accordingly the incidence of these four kinds of cancer was almost the same as that of Americans. This fact indicates that the site where cancer occurs may depend on the particular eating habits of the country of residence.

Table 2.5. Anticarcinogenic compounds contained in vegetables, fruits and spices

Foods	Compounds	Mechanism of anticarcinogenesis
Vegetables and fruits	Glutathione	Inhibition of formation of carcinogens
	Cysteine	"
	Gallic acid	Inactivation of carcinogens
	Caffeic acid	"
	Chlorogenic acid	"
	Quercetin	"
	Kempherol	Inhibition of activation of procarcinogens
Spices	Garlic	Induction of detoxication enzymes
	Onion	Suppression of promotion
	Saffron	"
	Tumeric	Suppression of promotion, induction of enzymes
	Curcumin	"
	Rosemary	Inhibition of binding between carcinogens and DNA
	Black pepper	"
	Japanese pepper	"

2. 2. 6. Anticarcinogens present in food materials

Recently many anticarcinogens have been found in our environment. The main anticarcinogens are shown in Tables 2. 5. and 2. 6.

While it has for a long time been thought that vegetables and fruits have anticarcinogenic activity, Kada *et al.* and Morita *et al.* in the National Institute of Genetics, Mishima, Japan, found that juices made from a selection of vegetables and fruits showed antimutagenic activities against mutations induced in *Salmonella typhimurium* by trptophan pyrolysis products. In particular vegetables such as cabbage, broccoli, eggplant, capsicum, ginger, horseradish, and burdock, and fruits such as apple, and pineapple had strong antimutagenic activity. Also, most spices such as garlic, onion, turmeric, saffron, rosemary, and natural spices such as cinnamaldehyde, vanilla, vanillin, coumarin, and vitamins such as vitamins C and E, medicinal plants and plant fibers have antimutagenic and anticarcinogenic activity.

Table 2.6. Anticarcinogens contained in natural spices, green tea and medicinal herb

Plants	Compounds	Part of Plant	Application
Natural spices	Cinnamaldehyde	Cinnamon bark	Foods, spicy cosmetics, medical supplies
	Cumarin	Tonnka beans, Kashia oil	Soap, lotion, tobacco, Deodorant
	*p*anisaldehyde		
	Vanillin	Vanilla beans	Chocolate,
	Vanillal	"	Ice cream, cakes
Green tea	Epicatechin, epicatechin gallate, epigallocatechin, epigallocatechin gallate, gallic acid, vitamin C, theaflavin, theaflavin gallate, theaflavin digallate		
Medicinal herbs	Monkey chair, peony (Paeonia lactiflora), Agrimonia pilosa, Rosa laevigate, Acanthopanax trifoliatus, Ardisia punctata, Kadsura coccinea		

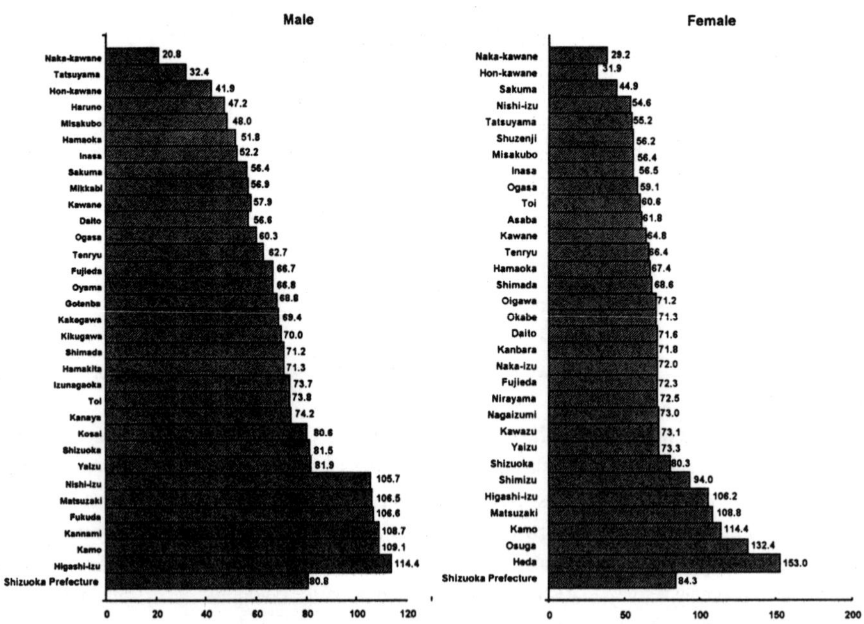

Figure 2. 9 A comparison of the death rate from stomach cancer in cities, towns and villages in Shizuoka Prefecture. (Oguni,, I., *et al.*,1989).

2. 2. 7. Tea drinking and stomach cancer

The first statistical study on the possible anticaricinogenic activity of tea was carried out in 1973, when population dynamic statistics were initially reported. In this report, the cancer mortality in Shizuoka was notably lower than the average rate of cancer mortality in Japan. Oguni, *et al.* (1989) in Shizuoka Junior College, found this result interesting and thus investigated the relationship between stomach cancer mortality and eating habits, in particular the amount of green tea consumed by residents of 75 cities, towns and villages in Shizuoka Prefecture.

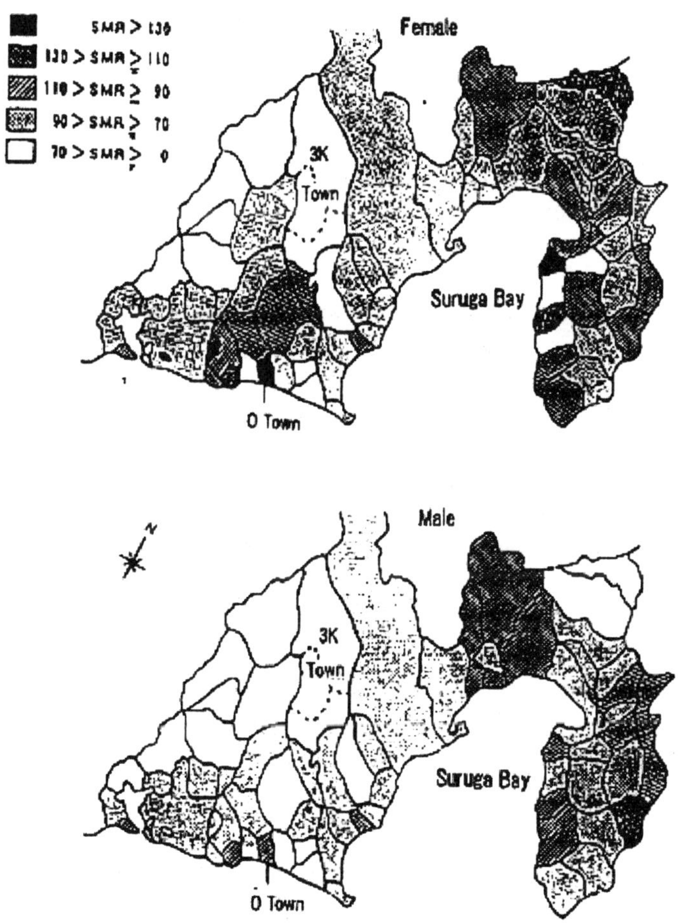

Figure 2. 10. The death rate of stomach cancer patients in each city, town and villege in Shizuoka Prefecture. (Oguni, I., *et al.*, 1989)

Figure 2. 9. indicates the stomach cancer mortality in males and females for each city, town and village in Shizuoka Prefecture. Figure 2. 10. shows the respective stomach cancer moralities for each different area on the map of Shizuoka Prefecture.

The average stomach cancer mortality rate in Shizuoka Prefecture was 80.8% in males and 84.3% in females, compared with 100% in the rest of Japan, indicating a lower level of mortality in Shizuoka Prefecture. Among the different districts in Shizuoka Prefecture, an area between the Abe River and Ohi River, in the center of Shizuoka Prefecture, in the upper area of the Tenryu River, and the North-East area in Shizuoka Prefecture showed extremely lower mortality rates for stomach cancer.

These areas are famous in Japan for their cultivation of green tea. The upper area of the Abe River produces a special brand, Motoyama-cha, and the upper area of the Ohi River produces another special brand, Kawane-cha. The area between the Abe River and the Ohi River is a famous tea producing area in Japan known as the three Kawanes: Hon-Kawane (Original Kawane), Naka-Kawane (Middle Kawane) and Kawane. On the other hand, Higashi-Izu (the East Izu) town, Matsuzaki town, Kamo village and Heda village show stomach cancer moralities five to ten percent higher than the average values for all of Japan.

Oguni, *et al.* conducted a survey on the frequency of drinking green tea for people between the ages of 35 and 70 in the three Kawanes (3K towns) where stomach cancer mortality is low and Osuga town (O town) where stomach cancer is high. The results obtained are shown in Figure 2. 11. To the question: "How frequently do you drink green tea? 1) almost not at all, 2) sometimes, 3) three times a day and 4) frequently in addition to three daily mealtimes, the number of people whose answer was 4),

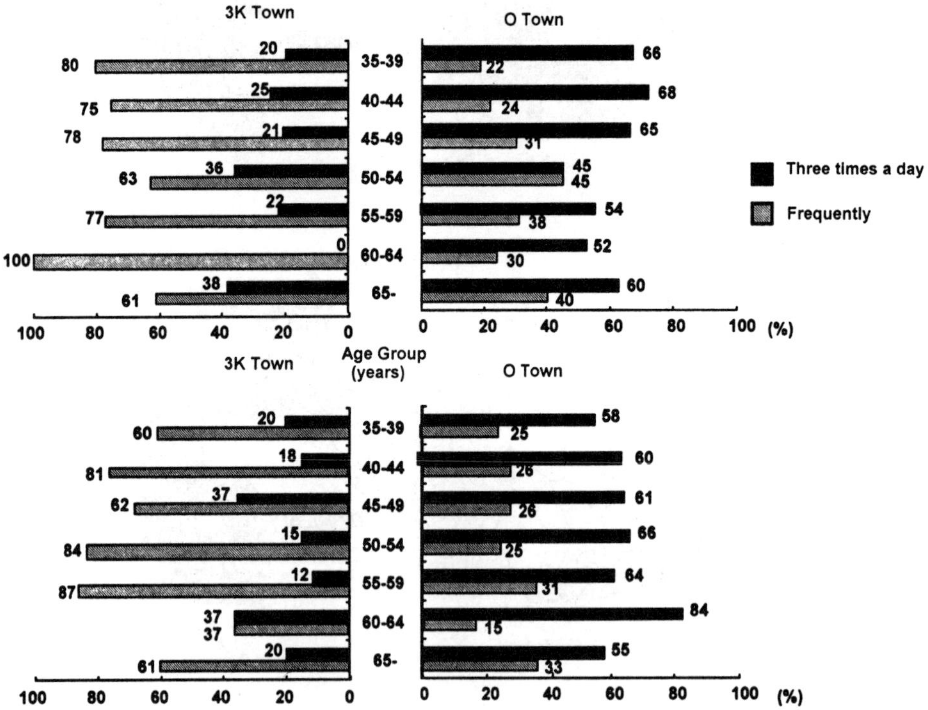

Figure 2. 11. Comparison of the frequency of tea drinking at meal times and between meals among residents in the 3 Kawanes and Osuga town in Shizuoka Prefecture. (Oguni, I., *et al.*, 1989)

was twice as high (with little difference according to gender or age) in the three K towns as in the O town. In particular, all men living in the three Kawanes, who were between the ages of 60 to 64, , drank green tea frequently at times other than mealtimes.

Furthermore, when making the tea, most of both the men and the women in the three K towns used fresh tea leaves whenever they drank tea, while most people in the O town used the tea leaves for several extractions, changing them only two to three times a day. People in the three K towns also

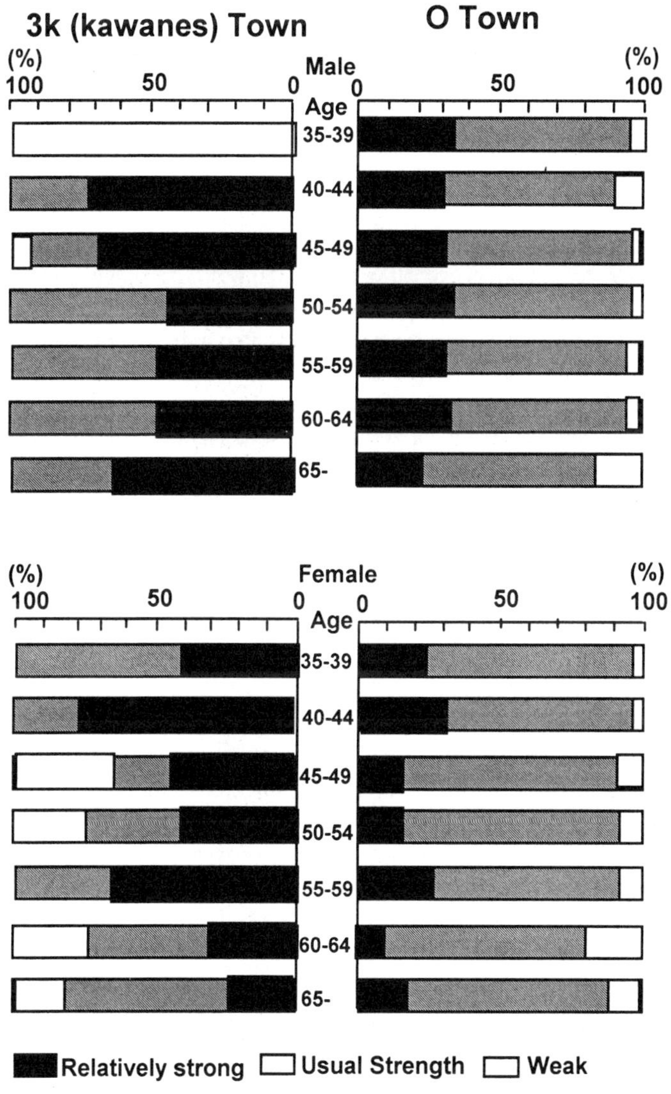

Figure 2. 12. Comparison of strength of tea drunk in the 3 Kawanes and Osuga towns in Shizuoka Prefecture. (Oguni, I, *et al.*, 1989)

drank strong tea, as shown in their response to the question: How strong do you make the green tea extraction? 1) relatively strong, 2) usual strength, 3) weak (Figure 2. 12).

2. 2. 8. Tea ceremony instructors and tea

There are many schools teaching Japanese tea ceremony (Cha-no-yu), including the Omote-Senke, the Ura-Senke, etc. The instructors of Japanese tea ceremony teach their students about the procedures (Otemae) and manners of preparing and drinking tea. Therefore, they have many opportunities to drink tea. The green tea used for the tea ceremony is grown covered in order to avoid direct sunshine, in a similar way to Gyokuro (the highest guality green tea) ; then it is steamed without rolling, spread out to dry, cut into small pieces and ground into powder with a stone mortar. In the tea ceremony, hot water is poured into the powdered tea, then it is mixed with a bamboo whisk (Chasen), and the whole beverage is drunk in its entirety. In this way it is not only the tea extract which is consumed, but also the tea leaves.

Sadakata *et al.* (1992) of Tohoku University Medical School examined the death rate and cause of death of 3.380 female tea instructors over 50 years old, who were teaching the Ura Senke style of tea ceremony and living in Tokyo. Data were examined for the 9 years from January 1, 1980 to December 31, 1988. When they compared the results obtained with those for a comparable group of females around the same age from the general public living in Tokyo, the number of tea instructors who died during this nine year period was 280, which was much less than the 439.9 from the general public and also about half the number of deaths for women of the same age throughout Japan (512.4).

When the causes of death of 167 of the deceased instructors was investigated by reviewing their death notices, it was found that 55 (32.2%) died from cancer, 41 (24.6%) died from cerebral hemorrhage, 36 (21.0%) from heart disease and 36 (21.6%) from other causes. These rates of death causes corresponded to those of the women of the general public living in Tokyo and those of women from throughout Japan.

2. 2 9. Antimutagenic activity of tea

In order to detect the carcinogenic substances (carcinogens) present in the environment efficiently and quickly, procedures can be employed which establish whether mutations of genes, DNA and chromosomes have occurred in the initiation process, and whether or not the second step of carcinogenesis is progressing. These procedures, by using bacteria, yeasts, fungi, insects and cultured mammalian cells to obtain results on carcinogenicity within a very short period, reduce the need for extensive, time consuming animal testing which can take two to three years.

Also, these short-term tests, can be used for detection of substances which inhibit the mutagenicity of mutagens and carcinogens in the initiation or promotion processes thus enableing identification of anti-carcinogenic substances in the environment. Kada, *et al.* (1985) discovered that green tea has an inhibiting activity against spontaneous mutations in *Bacillus subtilis*. Chemical analysis of the effective substances contained in green tea revealed four catechins, of which epigallocatechin gallate (EGCg) was found to have the strongest inhibiting action (Figure 2.13.).

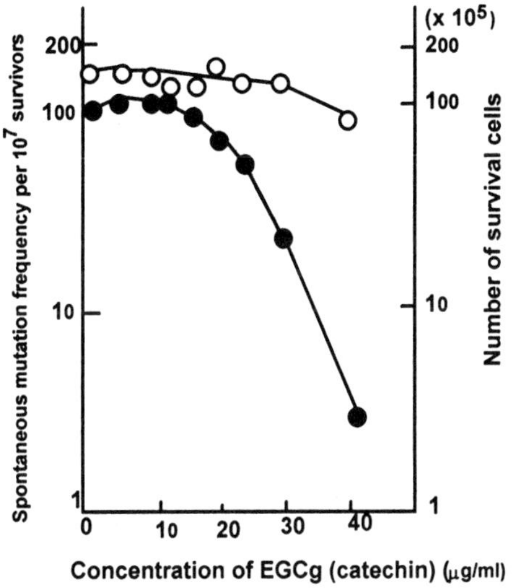

Figure 2. 13. Inhibition of spontaneous mutations by tea catechins in *Bacillus subtilis*. (Kada, T.,, *et al.* 1985).

What chemical substances are contained in tea? The chemical constituents found in tea are shown in Table 2. 1. The results indicate that 46% of the content is water insoluble hydrocarbons and 30% is food fibers. The protein content in tea is 24%, while the amount of water-soluble polyphenols, is 10 to 18%. Hot water-extracted green tea contains 15% catechins.

Almost all tea polyphenols are catechins, of which there are several different kinds with different molecular structures. In black tea, which is produced by fermenting the tea leaves, the same catechins which are contained originally in green tea react with each other during the process of fermentation, and tea polyphenols unique to black tea such as theaflavin, having an orange color, and thearubigin having a brown color, are produced. as shown in Table 2. 7.

Table 2.7. Polyphenol contents in various teas (%) (Muramatsu, K., ed., 1991)

Polyphenols	Green tea			Oolong tea	Black tea
	Gyokuro		Bancha		
Catechins	9.55	13.84	13.00	6.97	5.67
Theoflavins	—	—	—	0.10	1.07
Other polyphenols	2.40	1.76	0.41	9.04	14.60
Total	11.95	15.60	13.41	16.11	21.34

Among Japanese green teas, there are non-fermented green teas such as Gyokuro (high quality green tea), Sencha (powdered green tea), Matcha (powdered green tea), Kamairicha (pan-fired green tea), Bancha (later and coarse leaf green tea) and Hojicha (roasted green tea). There are also half-fermented oolong tea and full-fermented black tea. As shown in Table 2. 8, the amount of catechins contained in the everyday grade tea, Sencha, is greater than that contained in the higher-grade tea, Gyokuro. There is also a high content of catechins in Kamairicha, while not so much is contained in oolong and black teas.

Table 2.8 Comparison of contents in various teas (New Tea Industry Complete Works, Tea Industry Meeting Hall in Shizuoka Prefecture, 1988 ; Muramatsu, K., ed., Science of Tea, Asakura, Tokyo, 1991)

Teas	Total N (%)	Caffeine (%)	Catechins (%)					Teannin (mg/100g)	Amino acid (%)
			ECg	EC	EGCg	EGC	Total		
Gyokuro (High class)	6.31	4.04	1.35	0.36	6.65	1.68	10.04	2.65	5.36
Gyokuro (Middle class)	5.48	3.10	1.65	0.50	6.60	2.04	10.79	1.48	
Sencha (High class)	5.48	2.87	2.47	0.74	8.16	2.77	14.14	1.28	2.70
Sencha (Middle class)	5.35	2.80	1.76	0.91	7.53	3.36	13.56	1.21	2.18
Matcha (High class) (Konacha)	6.36	3.85						2.26	5.80
Matcha (Midde class) (Konacha)	5.85	3.52						1.79	4.61
Kamairi-cha (High class)	5.44	2.99	2.38	1.20	7.43	3.09	14.10		2.73
Kamairi-cha (Middle class)			2.68	0.86	10.85	3.77	18.16		
Bancha	3.83	2.02	2.40	1.07	5.58	3.28	12.33		0.77
Hojicha (Roasted tea)	3.46	1.93	1.56	0.40	5.00	1.36	8.32		0.20
Oolong tea	4.59	3.87	1.16	0.33	3.60	1.01	6.10		1.04
Houshucha (Paochung tea)	4.12	2.68	0.53	1.05	0.18	0.49	2.25		1.23
Black tea (Dajiling)			3.92	0.67	4.02		8.61		

The chemical structures of catechins and theaflavins are shown in Figure 2. 14. All these structures resemble each other closely.

Using procedures for detecting gene mutations and chromosome aberrations induced by mutagens and carcinogens in *S. typhimurium, E. coli, B. subtitis*, cultured cells of Chinese hamsters, and rats and humans, whether or not catechins and tea extracts inhibited or suppressed these gene mutations and chromosome aberrations induced by carcinogens was examined.

Results showed that gene mutations and chromosome aberrations induced by carcinogens such as tryptophan pyrolysis products, Trp-P-1 and Trp-P-2, benzo[a]pyrene, UV, 4-nitroqunoline-1-oxide (4NQO), nitrosoguanidine and aflatoxin B_1, were inhibited by tea catechins and extracts (Table 2. 9). Subsequently, it was found that the inhibitions of mutagenesis and carcinogenesis by tea catechins were due to the various activities of tea catechins in the processes during which normal cells mutate due to changes induced by carcinogens and mutagens.

2. 2 10. Anticarcinogenic activity of tea

As described previously, the multi-step carcinogenic process whereby normal cells become cancerous, consists of the following stages: "initiation", in which the induction of gene mutations occurs in the cell nucleus, "promotion", in which genes controlling the cell membrane and cytoplasm of the cells undergo changes, and "progression", in which there is stimulation of cell proliferation leading to the cells becoming cancerous. If tea suppresses the "initiation" or "promotion" or "progression", it may suppress the process of carcinogenesis.

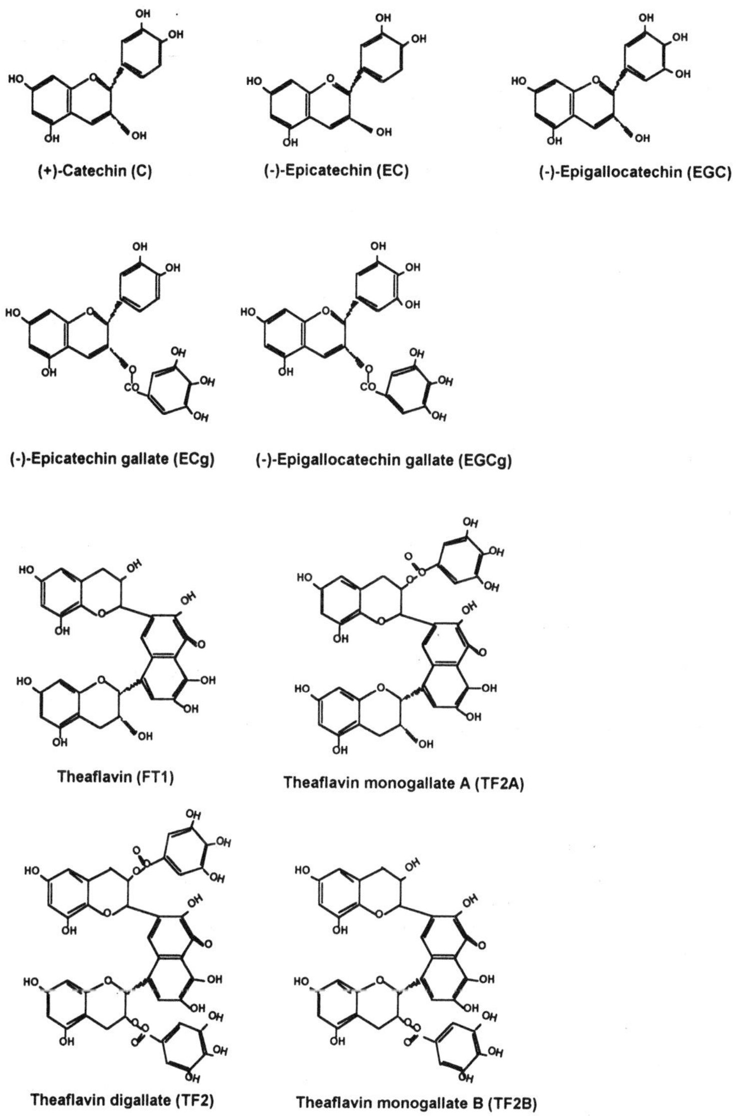

Figure 2. 14. Chemical structures of catechins and theaflavins contained in tea. (Kuroda, Y. and Hara,, Y., 1999)

Table 2.9. Inhibition by tea catechins of mutations induced by carcinogens

Test systems	Carcinogens and radiation	Tea catechins
Salmonella typhimurium TA98, TA100	Trp-P-2, N-OH-Trp-P-2	ECg, EGCg
	Benzo[a]pyrene	EGCg
	Nitrosoguanidine	Green tea extract
	Arylamine, heterocyclic amine	Polyphenol
	H_2O_2	Black tea, TF, TF2A, TF3
	Benzo[a]pyrene, aflatoxin B_1	Black tea extract
	Trp-P-1, Glu-P-1	
Escherichia coli B/r	UV, γ -rays	
	4NQO, nitrosoguanidine	Tannic acid
	UV, 4NQO, nitrosoguanidine	Green tea and black tea extract
Bacillus subtilis NIG 1125	Spontaneous mutations	EGCg
Chinese hamster cultured cells, V79	4NQO	ECg, EGCg
Rat bone marrow cells	Aflatoxin B_1	Green tea extract

Cancers were initiated in the esophagus, duodenum, fore-stomach, mammary gland, liver, lung or skin of experimental rats and mice, by exposure to carcinogens such as mitomycin C, nitrosoguanidine, aflatoxin B_1 or diethyl benzo[a]anthracene (DMBA). When tea or catechins were given to these animals, the onset of cancers in these tissues or organs was inhibited (Table 2. 10)

Table 2.10. Anticarcinogenic activity of tea catechins

Organs	Animals	Carcinogens	Tea catechins
Esophageal tumors	Rat	Methyl cholanthrene	EGCg
		Nitroso methyl benzyl amine	Decaffeinated green and black teas
Duodenal tumors	Mouse	Ethyl-nitro-nitrosoguanidine	EGCg
Colon cancer	Rat	Azomethymethane	Green tea polyphenols
Forestomach tumors	Rat, Mouse	Nitrosodiethylamine	Green tea infusion
Mammary gland tumors	Rat	Dimethyl benzoanthracene	Green tea catechins
Large intestinal tumors	Mouse	Dimethylhydrazine	Green tea catechins
Hepatomas	Mouse	Spontaneous	EGCg
Lung tumors	Mouse	Nitrosodimethyl amine	Green tea
Small intestinal tumors		Benzo[a]pyrene	
	Rat	Diethylnitrosoamine, etc.	Green tea catechins
Skin tumors	Mouse	Teleocidine	EGCg
		Methylbenzanthracene	TF, TF3, EGCg

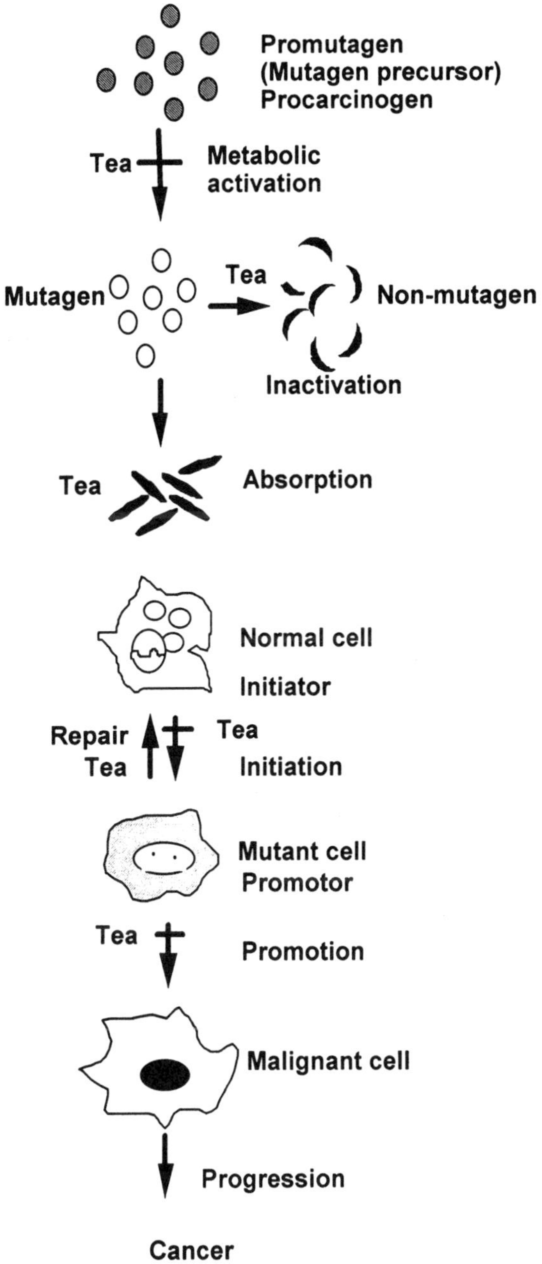

Figure 2. 15. Mechanisms of inhibition by tea of each step of carcinogenesis (Kuroda, Y., 1995).

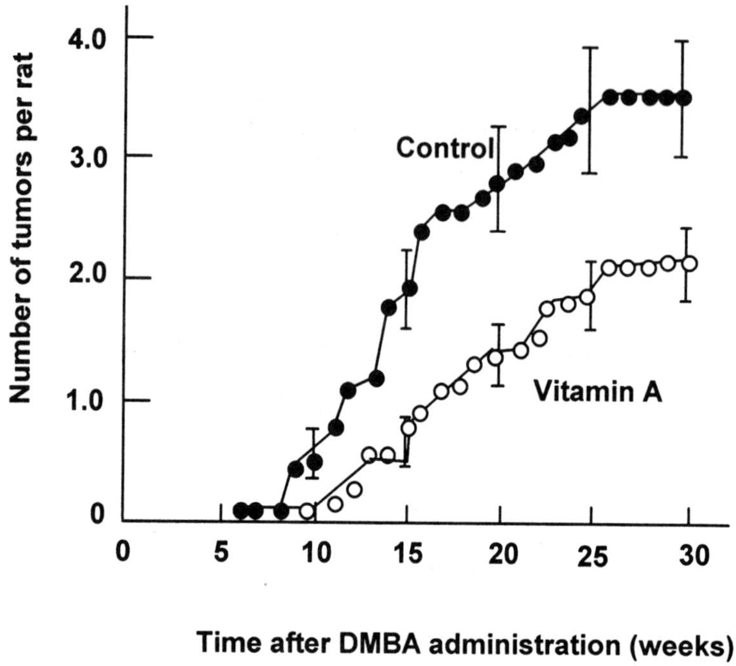

Figure 2. 16. Inhibitory effect of vitamin A on mammary tumors induced by DMBA (7,12-dimethylbenzo[a]anthracene) in rats. (McCormick, D. L., *et al.*, 1980)

The results indicate that almost all cancers are inhibited by drinking tea. It was also reported that human stomach cancer cells, hepatic cancer cells, and lymphatic leukemia cells *in vitro*, were destroyed by green tea extracts, EGCg or EGC.

Why did tea or polyphenols contained in tea suppress cancer? To answer this question, the mechanisms of cancer inhibition by tea extracts or polyphenols such as EGCg were examined. There were found to be many inhibition mechanisms of tea as shown in Figure 2. 15. The first mechanism of inhibition is the direct action of tea against carcinogenic substances, resulting in inactivation of carcinogens or binding of carcinogens to prevent their entrance into cells.

The second mechanism of cancer inhibition by tea is suppression of mutation induction by carcinogens after they have entered the cells, or induction of repair actions by tea against mutations even after mutations are induced by carcinogens. Other mechanisms are inhibition by tea of promotion and progression in the later processes of carcinogenesis. It is a surprising fact that tea polyphenols have so many inhibiting activities against the various steps which lead to the process of normal cells becoming cancerous. Recently, it was found that tea suppresses not only the development of normal cells into cancer cells, but also the spread of cancer cells from the original organs or tissues to other remote organs or tissues by metastasis. This metastasis of cancer cells occurs when the cancerous cells produce a proteolytic protein enzyme such as urokinase, which is secreted and breaks down the connections between the surrounding normal cells. Tea inactivates these proteolytic enzymes produced by cancer cells, resulting in prevention of cancer cell metastasis.

In addition to the anticarcinogenic activity as described above, tea also contains vitamins A, B and

E. Vitamin A is present as a yellow-green carotene similar to that found in some vegetables.. When we consume carotene, it is converted into vitamin A in the body. The lack of vitamin A can lead to a condition known as "night blindness", which is a disease that causes impairment of sight in the dark. In patients with this disease, the skin and the mucous membrane become weakened, and the resistance of the body declines making them more susceptible to colds. In experimental animals such as mice and rats, mammary gland tumors induced by dimethyl benzo[*a*]anthracene were suppressed by the consumption of vitamin A (Figure 2. 16).

Many believe that the most effective vitamin for cancer prevention is vitamin C, which is contained in lemon and radish juice, vegetables and abundantly in green tea. The lack of this vitamin weakens the resistance of the body and is the cause of scurvy disease. Recently it was found that vitamin C and its derivatives suppressed the initiation stage of cancers induced by ethyl methanesulfonate (EMS) in cultured Chinese hamster cells (Figure 2. 17). In experimental animals, skin, colon and lung cancers induced by carcinogens were suppressed by vitamin C.

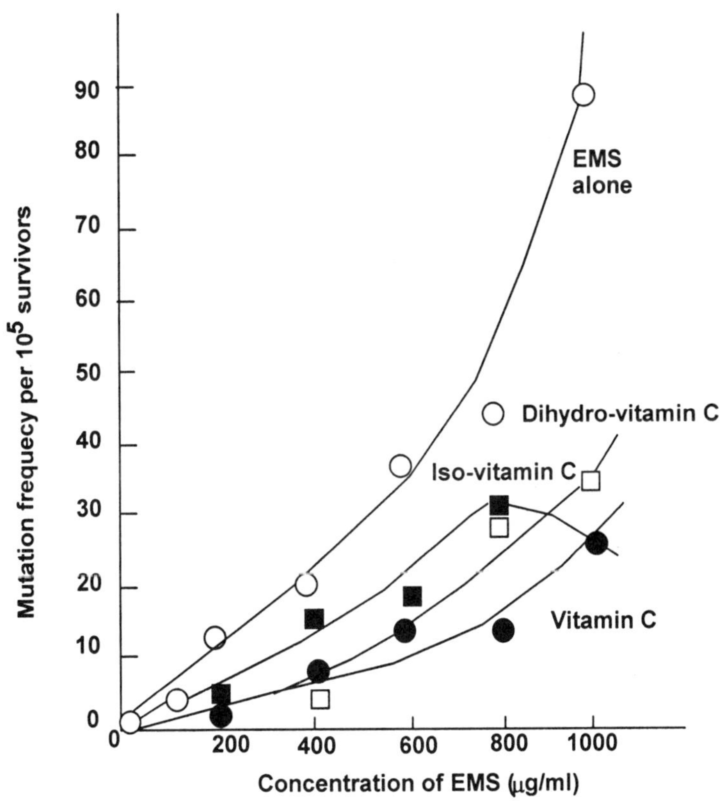

Figure 2. 17.. Inhibitory effects of vitamin C and its derivatives on mutations induced by EMS (ethyl methanesulfonate) in Chinese hamster cells. (Kuroda, Y., 1986).

The effect of vitamin C against cancer in humans was investigated by Cameron and Pouring. They injected 100g of vitamin C per day for 10 days, followed by 1g of vitamin C per day for 100 days, into 100 terminal cancer patients, and the results showed a higher survival rate of the patients who received Vitamin C.

Vitamin E is found in large amounts in many foods, especially in sesame corn oil and soybean oils, and is also contained in adrenal glands, ovaries, liver and pancreas in the animal body. Vitamin E is abundant in tea leaves, but it is not present in the tea extraction due to its insolubility to water. It has a strong antioxidant activity and prevents the process of "promotion", the second step in carcinogenesis. Because of its insolubility in water, to reap the health benefits of its presence in tea, it is necessary to consume the tea leaves. Drinking the tea of the Japanese tea ceremony is an excellent way of making the most of the vitamin E in tea, since the tea used is Matcha, which is powdered green tea that is extracted and drunk in its entirety.

2.2.11. Programming cell death (Apoptosis) induced by tea

Programming cell death (Apoptosis) is a common mechanism of morphogenesis in normal development of multi-cellular organisms. Disappearance of the tadpole tail in metamorphosis of the frog and cell death in webs in the finger formation in hands and feet are examples of the apoptosis phenomena in normal development of higher animals.

It was found by Hibasami, *et al.* (1996) that apoptosis was induced by tea catechins, EGC and EGCg, in human lymphoid leukemia Molt 4B cells. Nakamura, *et al.* (2002) isolated a water-soluble high molecular weight fraction from green tea (tea non-dialysate, TND). This fraction was a mixture of sugars and polyphenols at concentrations of 0.42% in green tea, 2.01% in black tea and 1.78% in oolong tea. The fraction inhibited the promotion process in carcinogenesis of mouse epidermal cells JB6 and induced apoptosis (Nakamura, 2002). The TND fraction was also effective in inducing apoptosis in human lymphoma, stomach and colon cancer cells at a concentration non-toxic to normal human cells (Hayakawa, *et al.*, 2001).

Yang, *et al.* (2002) showed that chemically induced tumors were reduced in volume and multiplicity by 50% by treatment with 0.6% green tea as the sole source of drinking fluid in the hamster. In addition, immunohistochemistry showed that tea induced the apoptosis (Li, *et al.*, 2002). Using a mouse model of prostate carcinogenesis, it was recently demonstrated that oral consumption of 0.1% green tea polyphenols decreased tumor incidence by 65%. In contrast to water-drinking control animals, which had a high rate of distal metastasis (25-95%), tea-drinking mice showed a significant decrease in proliferating cell number and a 10-fold increase in tumor cell apoptosis (Gupra, *et al.*, 2001).

The enzyme digestion product of green tea residue, which contained cellulose, pectinase and proteases was also found to inhibit cell proliferation of human histocytic lymphoma cells by inducing apoptosis (Katsuno and Isemura, 2002).

Isemura (2002) found that EGCg bound to cell surface death receptor Fas to trigger the apoptosis caused by activation of caspase 8 enzyme in the cytoplasmic caspase cascade. These findings indicated that tea catechins selectively induced the apoptosis of cancer cells, resulting in their death without any toxic effect to normal cells, and produced the prevention of tumor formation.

2. 2. 12. Inhibition of angiogenesis by tea

Angiogenesis has been revealed to be a critical process for solid tumor growth and metastasis. Thus, inhibition of this process causes suppression of tumor growth by cutting off the supply of oxygen and nutrition. Angiogenesis is initiated by the release of some factors from tumor cells, such as basic fibroblast growth factor (bFGF) and vascular endothelial growth factor (VEGF). These factors activate the endothelial cells of the blood vessels and promote the migration and growth of these cells.

Kondo, *et al.*, (2002) and Oku, *et al.*, (2003) found that tea catechins, EC, ECg, EGC and EGCg had inhibitory activity against growth, migration and vessel tube formation in *in vitro* angiogenesis model system using human umbilical vein endothelial cells (HUVEC), and *in vivo* angiogenesis system using tumor cells implanted in dorsal sac model mice (Figure 2. 18).

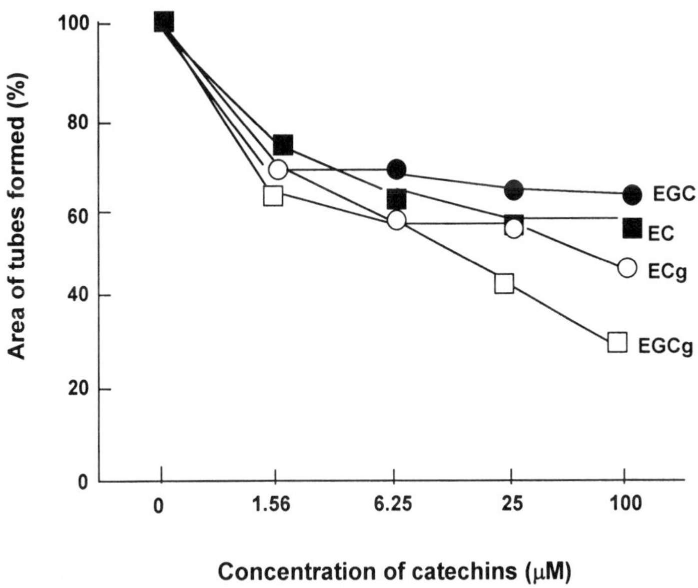

Figure 2. 18. Inhibitory effects of tea catechins on the tube formation by HUVEC cells in three dimensional cultures with collagen gel (Kondo, T., *et al.*, 2002))

2. 2. 13. Inhibition of *Helicobacter pylori* growth by tea

Chronic inflammation induced by *Helicobacter pylori* is associated with a causative pathway in carcinogenesis of stomach tumors. The anti-inflammatory effects of tea catechins in gastric epithelial cells were examined by Handa, *et al.* (2002). In *H. pylori*-associated gastric inflammation, interleukin-8 (IL-8) is produced by gastric epithelial cells and directs neutrophils to the gastric mucosa. The production of IL-8 was inhibited by tea catechins in HUVEC cells (Honda, *et al*, 2001). This indicates that tea catechins inhibit carcinogenesis pathway through the anti-inflammatory effects in the gastric mucosal inflammation.

2. 2. 14. Clinical chemopreventive trials with tea catechins

A lot of evidence related to the preventive and inhibitory activity of tea catechins against carcinogenesis has been accumulated by using *in vitro* cell systems and *in vivo* animal experiments. However, the final goal of the confirmation of the preventive effects of tea catechins against carcinogenesis should be to confirm their effects on human beings. Some epidemiological data suggested the beneficial effects of tea catechins against human carcinogenesis, as described previously.

Some clinical and chemopreventive trials with tea catechins have been underway in a cooperative study between the National Cancer Institute in USA and Tokyo Food Techno., Co., Ltd; in Japan (Hara, 2003). "Polyphenon E", which contains 60% EGCg and 30% other tea catechins, were used as the standard sample in this trials. Studies on the safety of the sample were carried out with the tests on the mutagenicity in microorganisms, the genotoxicity,and the embryonic toxicity in higher animals, and pharmacokinetic studies in dog and human. In the *in vivo* kinetic study using dogs, it was shown that H^3-labeled EGCg was absorbed through the digestive tract, distributed in various tissues and organs, and excreted outside of the body.

In the study on safety for humans (clinical first phase experiment) , after a single intake of tea catechins, the amounts of catechins and their metabolites in the blood and urine were determined on a time course. A pharmacokinetic study was also conducted where tea catechins were taken continuously for 28 days, on both an empty and full stomach. The study of the effects on humans (clinical second phase experiment) is now being carried out on precancerous states of skin, bladder, lung and other various organs and tissues in several universities and institutes in USA. It is expected that results showing benefits and the preventive effects of tea catechins against human carcinogenesis, will be obtained in the near future.

2. 3. Prevention against allergy

When the human body was sensitized previously by an antigen and there is exposure to the same antigen again, the body responds rapidly to this antigen. Of these responses, the beneficial and protective response in human body is known as "immune response". On the other hand, the harmful response in the body is known as "allergy".

The antigens which produce allergic response (called allergens) include house dusts, ticks, and pollen of cryptomeria and cedars. Recently many people in Japan suffer from allergies mainly caused by cedar pollens every spring. They are plagued by an allergic conjunctivitis and nasal allergy. In the process of allergic response, platelet-activating factor (PAF) acts as a potent chemical mediater and activates numerous cell types such as macrophages, mast cells and endothelial cells.

Sugatani, *et al.,* (2003) found that tea catechins, ECg, EGCg、theaflavins inhibited platelet aggregation by PAF, suggesting the prevention of allergic reaction.

APPENDIX 2

COLOR, TASTE AND AROMA

Tea in its original state is green tea, which becomes oolong or black tea by oxidative fermentation. Catechins contained in green tea are oxidized by the polyphenol oxidizing enzyme present in tea leaves, forming theaflavins and thearubigin. The activity of the polyphenol oxidizing enzyme varies depending on the varieties of tea tree, and the season. The activity of the enzyme is stronger in the Assam variety of tea tree and in trees grown during the dry season. Black tea produced from tea leaves picked in this season has good quality.

COLOR OF TEA

When hot water is poured onto tea leaves, the compounds contained in the tea leaves are extracted and color appears. In the case of green tea, the color is reddish gold. When green tea is extracted for a long time, the color changes to a reddish one. Thus, when green is taken for example in a thermos to a picnic, the tea extract turns reddish brown in color and moreover the taste and aroma of the tea deteriorate.

TASTE OF TEA

Green tea, oolong tea and black tea all have their own distinctive flavors. The taste of green tea is especially delicious, and may come from the catechins and amino acids. Its taste is at first bitter and astringent and then sweet. Gyokuro and the finest quality Sencha teas have an excellent balance of sweet and bitter astringent tastes and the overall flavor is strong. In the middle and lower quality Sencha teas, the bitter a stringent taste is three times stronger than the sweet taste. These differences in taste are not due to catechin content but to amino acid content. Black tea and green tea are different in flavor due to the fermentation process during production of black tea, in which catechins are oxidized and polymerized to form theaflavins ; and consequently the astringent taste of green tea is replaced with a more complex astringent taste. Further, the additional caffeine produces a deep bitter and astringent taste.

AROMA OF TEA

Tea has a unique fresh aroma which is at the same time relaxing. Aromas of compounds differ according to the variety of tea, and more than three hundred different aroma compounds are found in tea. While the fresh tea leaves have a somewhat grassy-smell, this lessens after steaming in the process of green tea production, and the refreshing fragrances of young leaves and flowers similar to lily of the valley and rose are apparent.

3

AGE-RELATED DISEASES AND TEA

3. 1. PREVENTION OF HYPERTENSION

3. 1. 1. Cause of hypertension

In the human body, the blood circulates to distribute oxygen taken into the lungs by respiration to the organs and to collect and bring back to the lungs carbon dioxide accumulated in each organ. Nutrients absorbed in the stomach and intestine are also conveyed by the blood to all organs and tissues. In a human body weighing 50 to 60 kg, blood has a volume of one twelfth to one thirteenth of body weight, that is 4 to 5 liters. It a circulates from the heart, by way of the blood vessels bringing oxygen and nutrients to all organs in the body. It is the pressure the blood exerts on the blood vessels during this process which is referred to as blood pressure.

The blood pressure is usually measured by a tonometer. When the blood pressure value is at its highest, as the heart is contracting to pump blood, this is known as "maximum blood pressure" (or systolic), and when the blood pressure value is at its lowest as the heart relaxes, this is known as "minimum blood pressure" (or diastolic).

During exercise, our heart beat increases in order to pump the blood faster, and as a result blood pressure rises. On the other hand, when we are lying down resting our blood pressure goes down. Blood pressure also changes at different times during the day, according to whether it is morning, the middle of the day or nighttime; and due to the differences in temperature in summer or winter. As we age, our blood vessels lose their elasticity and harden, resulting in higher blood pressure. If an elderly person goes into a cold bathroom and suddenly pours hot water over themselves, their blood pressure rises rapidly and this can result in heart attacks or cerebral hemorrhages.

High blood pressure, when the maximum blood pressure is continuously higher than 160 mm Hg or the minimum blood pressure is continuously higher than 95 mm Hg, generally indicates an unhealthy condition of the body. Outward symptoms include dizziness, stiff shoulders, and insomnia, and irritation; but many people fail to notice that anything is wrong until their blood pressure increases considerably. Secondary high blood pressure can be caused by other diseases, but more than 90% of high blood pressure is called essential high blood pressure and is not accompanied by any other particular disease. What causes essential high blood pressure? When nutrients are lacking in the diet and salt intake increases, the balance of nutrition is disturbed, leading to accumulation of cholesterol in the blood vessels which causes the blood vessel walls to thicken thus obstructing the flow of blood.

Particularly in the case of essential hypertension, Angiotensinogen contained in the blood is converted to Angiotensin I by rennin, an enzyme produced in the kidneys. Then Angiotensin I is converted to Angiotensin II by the enzyme Angiotensin I converting enzyme (ACE). Angiotensin II has a strong vasoconstricting action in the microcirculation system, resulting in an increase in the blood pressure (Figure 3. 1).

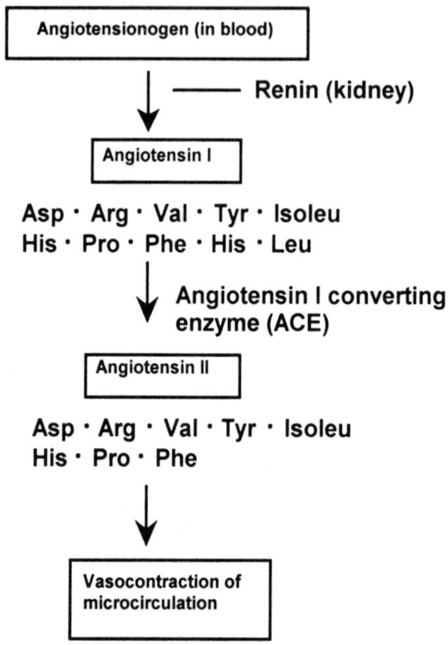

Figure 3. 1. Biosynthesis of angiotensin II, which causes the contraction of blood vessels, resulting in essential hypertension disease.

The ACE enzyme was extracted from pig kidneys and when various tea polyphenols were added to the ACE solution, the enzyme activity was strongly suppressed by tea polyphenols. This indicates that tea polyphenols can suppress vasoconstriction by inhibiting the ACE enzyme. Extracts of soybeans, shellfish, fruits and buckwheat also have an inhibiting activity on the ACE enzyme.

3. 1. 2. Hypotensive action of tea drinking in animal experiments

A special strain of genetically bred spontaneously hypertensive rats (SHR) develop hypertension as they grow. Rats in this strain show a hypertension rate as high as 100%, and are used as a model for human hypertension.

Five week old rats of this SHR strain were divided into two groups. The first group received a normal diet, while the second group was given a diet containing 0.5% of the green tea catechin EGCg.

The blood pressures of rats in both groups were taken every two weeks. As shown in Figure 3. 2 the blood pressure of rats in both groups was 130-140 mm Hg at the beginning of the experiment. The blood pressures of rats reared on the normal diet increased rapidly and exceeded 210 mm Hg at the age of 15 weeks. On the other hand, the blood pressures of rats reared on a diet containing catechins was suppressed to a value less than 200 mm Hg.

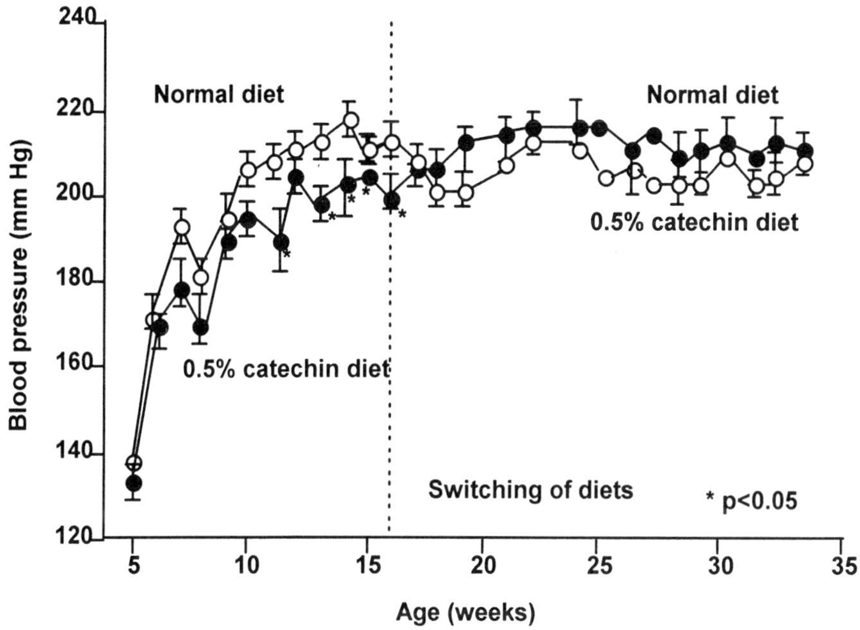

Figure 3. 2. Suppression of elevation of blood pressure in spontaneously hypersensitive rats (SHR) administered crude catechins . (Hara, Y. and Tono-oka, F.,, 1990).

When the diet of each group was switched to the other diet at the age of 15 weeks, the blood pressures of rats in both groups was reversed; that is, the high blood pressure of rats previously reared on the normal diet became lower than those of the rats with low blood pressure previously reared on the catechin diet. These results reveal that even rats that have a genetic tendency to hypertension can have their blood pressure suppressed when they are reared on a diet containing catechins. When these rats were put back on a normal diet, their blood pressures rose again. This shows that tea catechin has a hypotensive effect on blood pressure.

Many rats of the strain SHR die due to brain strokes. After repeated crossbreeding within this strain, a new strain was established with a death rate of more than 95% from brain strokes. These stroke-prone SHR rats (SHRSP) were again divided into two groups. The first group received a normal diet, while the second group was given an 0.5% catechin diet. Results indicated that there was no rise in the blood pressures of rats in the SHRSP strain reared on the catechin diet.

Rats reared on the normal diet showed symptoms of brain stroke from about week 15, and started dying one after another. By the 22nd week almost all of the rats had died. On the other hand, the first death in the group of rats reared on the catechin diet occurred 3 weeks later than that of the rats on the normal diet, and their survival time was 15% longer than those reared on the normal diet (Figure 3.3).

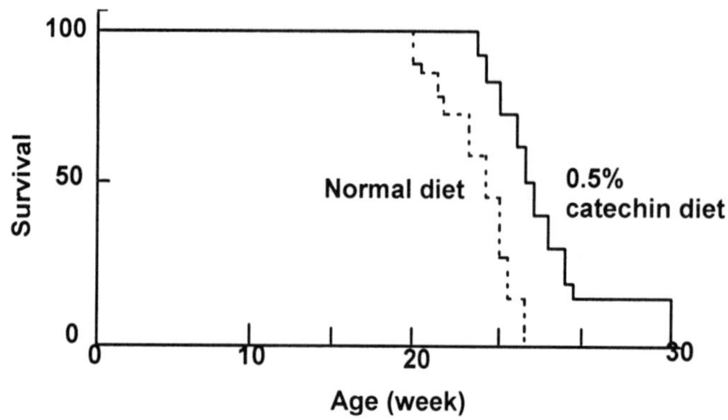

Figure 3. 3. Extension of life span in SHR by administration of crude catechins (Hara,Y. and Tono-oka, F.,. 1990).

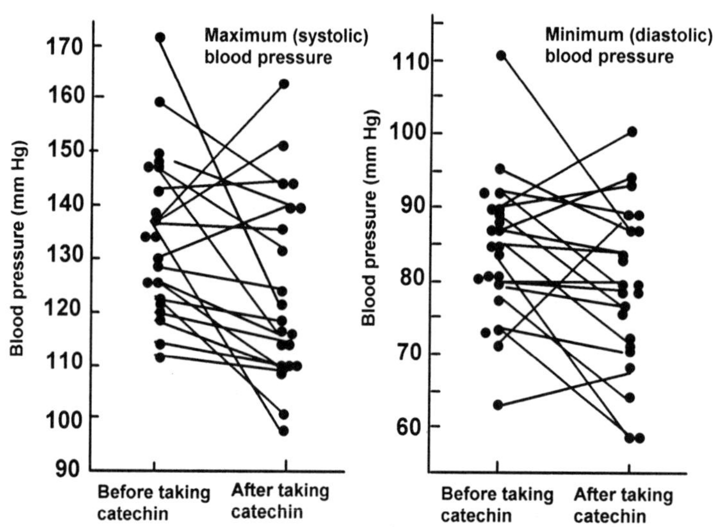

Figure 3. 4. Effect of catechin administration on human blood pressure. (Kanaya, S., *et al.*, 1991).

3. 1. 3. Hypotensive action of tea drinking in humans

Do tea catechins also show similar hypotensive activity on blood pressure in humans to what they do in rats? Under medical supervision, adult males (ages 20 to 70 years old) were given about 400 mg a day of tea catechin (Mitsui Norin Co., Polyphenon) for three months. Their blood pressures were determined before and after taking catechins.

As shown in Figure 3. 4, both the maximum (systolic) and the minimum (diastolic) blood pressures were significantly lowered by taking tea catechin. After taking tea catechin, there was a surprising drop in the blood pressure of a hypertensive person with a high systolic blood pressure of 170 mm. These results indicate that tea catechins are also effective in lowering human blood pressure.

3. 2. PREVENTION OF DIABETES

3. 2. 1. Cause of diabetes

Along with cancer and hypertension, diabetes is one of those afflictions which is more prone to strike as we get older. It used to be regarded as a disease of affluence in Japan since it was prevalent among rich people who indulged in a diet of delicacies. But due to the economic boom, there was a marked rise in the standard of living throughout Japan, and many people came to be able to afford more luxurious eating habits, which lead accordingly to an increase in the number of diabetes patients.

Most diabetes patients have no apparent symptoms early in their lives, and the typical case may be someone older than middle age who does not exercise regularly and is overweight. Very often these people have no personal history of disease and are proud of the fact that they hardly ever need to visit the doctor. The first sign that something is amiss is usually when they have a periodic health check and blood test results show high blood sugar level. The number of diabetes patients in Japan is more than two million. If we add to this the people who are still unaware they are diabetic, this number is likely to double.

If diabetes is just left to run its course and no action is taken to halt it, the patient is susceptible to arteriosclerosis or myocardial infarction. High blood sugar levels may also be the cause of damage to the capillary vessels in the eye, leading to bleeding in the eye and in the worst case, to blindness. Sugar is a carbohydrate which is one of the three nutrients that is necessary for life, the other two being protein and lipid; and carbohydrates are obtained from rice, bread, potatoes, etc. Polysaccharides ingested in the form of starch or complex sugars are digested by amylase excreted from the salivary gland or pancreas, and by maltase or sucrase secreted from the small intestine, to produce glucose, which is then absorbed through the small intestine. Glucose absorbed is transferred into the blood vessels and circulates to the whole body in the blood, being distributed to various tissues and organs for use as an energy source.

3..2. 2. Insulin and sugar-resolving enzymes

Glucose circulating in the blood vessels requires insulin, a hormone secreted from the pancreas, in order to be absorbed into tissues or organs. Diabetes occurs when amounts of insulin decrease due to insufficient secretion. Glucose is thus unable to be absorbed into the organs so glucose concentrations in the blood increase, and large amounts are excreted in the urine.

Diabetes patients must first restrict their intake of rice or bread, which contain starch and other carbohydrates. More advanced patients require supplemental insulin by injection. Starch and carbohydrates cannot be absorbed into the small intestine without being resolved into glucose. Inhibition of the resolution of starch or sugar results in the suppression of their absorption into the small intestine. Based on this mechanism, the drug "Acapose", produced by *Actinomyces,* was developed as a treatment drug for diabetes.

Table 3.1. Inhibition of the activity of α-amylase by tea catechins (Hara, Y. and Honda, M., 1990)

Tea catechins	Amounts for 50% inhibition (μM)
(-)-Epicatechin (EC)	>1,000
(+)-Catechin (C)	>1,000
(-)-Epigallocatechin (EGC)	>1,000
(+)-Gallocatechin (GC)	>1,000
(-)-Epicatechin gallate (ECg)	130
(+)-Catechin gallate (G)	20
(-)-Epigallocatechin gallate (EGCg)	260
(+)-Gallocatechin gallate (GCg)	55
Theaflavin (TF)	18
Theaflavin gallate A (TF2A)	1
Theaflavin gallate B (TF2B)	1.7
Theaflavin digallate (TF3)	0.6
Gallic acid	>1,000

Table 3.2. Inhibition by tea catechins of the sucrase activity in the intestines of rats (Honda, M. and Hara, Y., 1993)

Tea catechins	Concentration (mM)	Sucrase activity (%)
Control (no addition)	0.5	100
Gallic acid	0.5	93
Catechin (C)	0.5	86
Epicatechin (EC)	0.5	85
Epigallocatechin (EGC)	0.5	83
Epicatechin gallate (ECg)	0.5	38
Epigallocatechin gallate (EGCg)	0.5	21
Theaflavin (TF)	0.1	95
Theaflavin gallate A (TF2A)	0.1	81
Theaflavin gallate B (TF2B)	0.1	87
Theaflavin digallate (TF3)	0.1	58

Recently it was found that catechins contained in green tea, EC, EGC, ECg and EGCg, and theaflavins contained in black tea, TF1, TF2A, TF2B and TF3, have a strong inhibiting action on the sugar-resolving enzymes α-amylase and sucrase. In these studies, α-amylase obtained from human saliva was mixed with starch solution and incubated at 37°C for 10 min, and the amount of maltose resolved was determined. Results are shown in Table 3. 1. Catechins in tea had marked inhibitory effects on a-amylase activity. ECg and EGCg had especially strong inhibiting activity on α-amylase.

In the case of sucrase prepared from rat intestine membrane, the amount of glucose produced when treated with sucrase at 37°C for 10 min, is shown in Table 3. 2.

The inhibiting activity of catechins or theaflavins for sucrase was found in ECg and EGCg in green tea and theaflavins in black tea.

3. 2. 3. Tea lowers blood sugar levels in animals

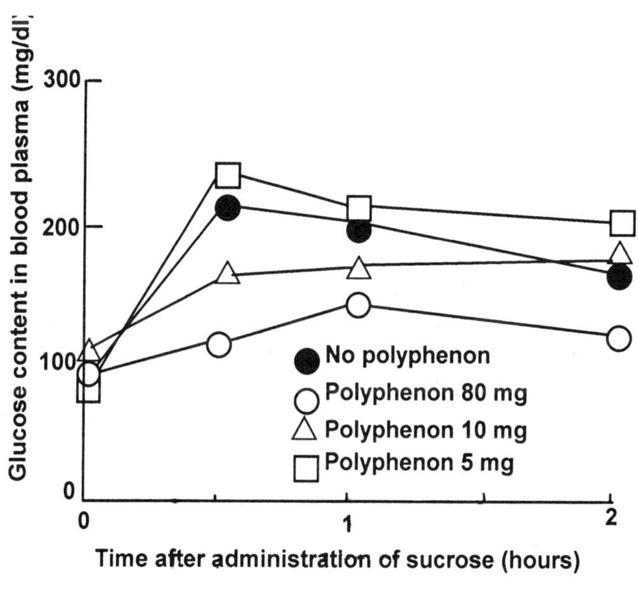

Figure 3. 5. Ef blood sugar levels induced by sta

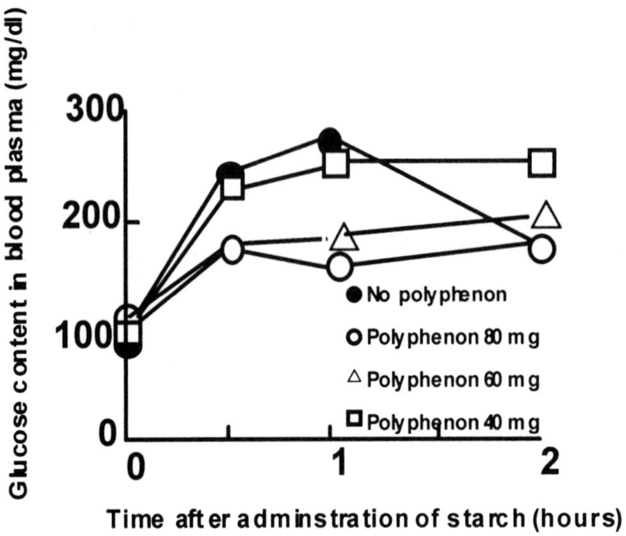

Figure 3. 6. Effect of tea catechins on suppression of an elevation in blood sugar Level induced by sucrose in rats . (Matsumoto, N., *et al.*, 1991)

Since tea was found to have an inhibiting action on sugar resolving enzymes, the possibility that tea might decrease the blood sugar levels in experimental animals was examined.

Rats 5 weeks old, reared on a normal diet, were given 4 ml of 40% soluble starch or sucrose after fasting overnight. Some rats were given 1 ml of green tea catechin solution before being given these sugars. Thirty minutes, 1 hour and 2 hours after sugar was given, blood was collected from the rats and the amount of glucose in the blood was determined. For rats given starch or sucrose, in each experiment the level of blood sugar was decreased by giving tea catechins (Figures. 3. 5, 3. 6)

3. 2. 4. Tea lowers human blood sugar levels

It was found that tea lowered human blood sugar levels. Under the supervision of medical doctors, 50 g of starch or sucrose were given to healthy volunteers. Their blood was collected and the level of blood sugar was determined before taking sugar, and 30 min, 1 hour, 2 hours and 3 hours after taking the sugar. To determine the effect of tea, tea catechins (polyphenon, 500 mg) were given before sugar intake and changes in the blood sugar levels were determined. Results are shown in Figure 3. 7. It was found that tea catechins suppressed the increase of blood sugar levels caused by consumption of starch or sucrose.

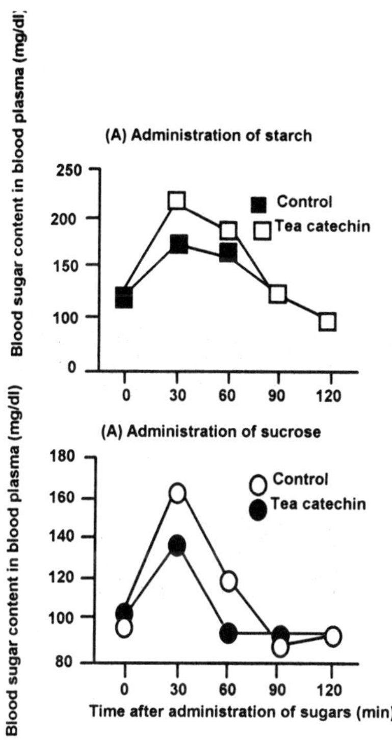

Figure 3. 7. Effect of tea catechins on suppression of an elevation in blood sugar level in human after administration of starch (A) or sucrose (B). (Sakai, M. and Hara, Y., 1995).

Tea catechins have no inhibitory effect on lipase which resolves lipids, or on trypsin which resolves proteins. However, tea catechins are effective in inhibiting the activity of amylase and sucrase, which resolve saccharides such as starch and sucrose. Tea drinking is effective in decreasing blood sugar levels without any effect on digestion of lipids and proteins. Therefore, tea may be safe for human health and an effective drink for the reduction of diabetes and obesity.

3. 3. PREVENTION OF OBESITY

Tea inhibits the resolution of sugars and prevents the absorption of resolved glucose in the digestive tracts, resulting in a decrease in the level of glucose in tissues and organs in the human body. As a result, the amount of sugar accumulated in the body decreases, followed by a decrease in lipids produced from sugar. This is an indication that tea decreases the amount of lipids in the body, which in turn inhibits weight gain.

Tea not only suppresses weight gain by affecting sugar absorption, but also suppresses the accumulation of lipids in the animal body or liver resulting from consuming lipids in foods. An experiment was carried out using male mice, 13 weeks old, which were divided into 3 groups of 8 mice.

Mice in the first group were fed coconut oil in food. Mice in the second group were fed a mixture of coconut oil and 0.1% green tea catechins. Mice in the third group were fed a normal diet. Mice in the three groups were later killed and the amounts of fat in their body and liver were determined.

Results are shown in Figure 3. 8. No differences were found in the body weights or in the amounts of food taken in by mice in the three groups. Although the amount of fat in the liver increased in the mice fed on the coconut oil diet as compared with mice fed on the normal diet, the amount of fat in mice on the coconut oil diet that were fed catechins dropped down to the level found in rats fed the normal diet.

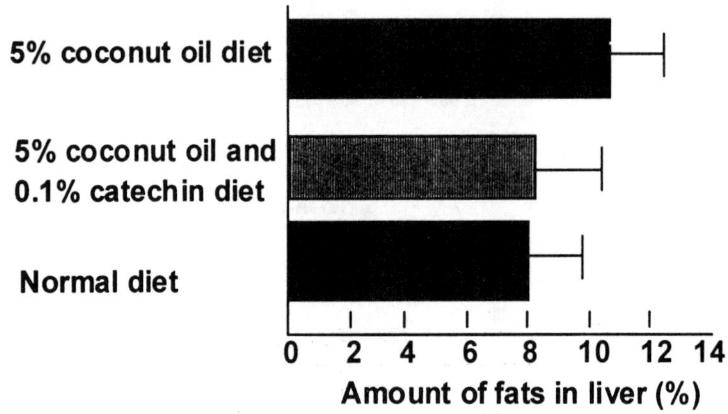

Figure 3. 8. Effect of tea catechins on suppression of an increase of liver fat after administration of coconut oil to mice. (Ishigaki, A., *et al.*, 1991).

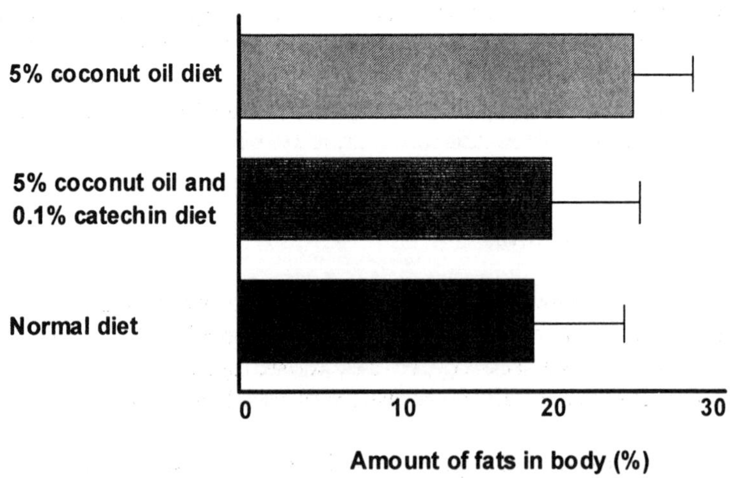

Figure 3. 9. Effect of tea catechins on suppression of an increase of body fat after administration of coconut oil to mice. (Ishigaki, A., *et al.*, 1991).

The amounts of fat in their bodies also increased in rats fed coconut oil as compared with rats fed on the normal diet, while the amounts in rats fed catechins dropped down to the normal level (Figure 3. 9).

The amount of 0.1% green tea catechins used in these experiments corresponds to the amount of catechins contained in 5-10 cups of green tea, which is consumed commonly in daily life by many Japanese people.

People who are in the habit of drinking this amount of green tea every day, may have suppressed accumulation of fat in the body and liver. Thus tea catechins are regarded as having the ability to inhibit the accumulation of fats produced from sugars and lipids in the diet.

3. 4. SUPPRESSION OF ARTERIOSCLEROSIS

3. 4. 1. Mechanisms of arteriosclerosis

It is said that aging begins in the blood vessels. Arteriosclerosis that occurs in the blood vessels may be one of the main causes of aging. As people age, lipids and plaque accumulate gradually on the walls of the blood vessels, and the resulting rough inner surfaces cause them to lose their elasticity and thicken and narrow. As stated before, the first cause of death in Japan is cancer, followed by heart disease as the second cause and cerebral apoplexy as the third. These second and third causes are due to blockage in the blood vessels, which makes them fragile and susceptible to tearing leading to a stoppage in the flow of blood to the heart; and to brain hemorrhage leading to damage to the brain.

The causes of arteriosclerosis, in addition to high blood pressure, are closely related to the amount of cholesterol contained in blood. Cholesterol is a substance common to all animals. Two thirds of the amount required is synthesized in the liver and small intestine, while the remaining one third is obtained from consumption of animal lipids such as meat, eggyolk, butter, etc.

3. 4. 2. What is cholesterol?

Cholesterol is an essential compound in the human body due to the important role it plays. Bile acid, which is required for digesting the lipids of foods, and hormones that are responsible for an amazing number of activities in the body, are made from cholesterol.

A large amount of cholesterol is also contained in the cellular membranes of all organs and in tissues such as liver, lung, kidney, etc. in the body, as well as in the intercellular matrix connecting cells in these organs.

Why is it that cholesterol, which is so vital to the body, can be the cause of arteriosclerosis? Cholesterol is a kind of lipid that is insoluble in water. When the blood containing cholesterol circulates in the body, the cholesterol binds to proteins and forms lipoproteins. These lipoproteins wrap cholesterol inside while the outside is covered with proteins, thus making them soluble in water. These lipoproteins are classified into three groups, depending on the kind of lipids inside and the kind of proteins outside.

3. 4. 3. Good cholesterol and bad cholesterol in the blood

The various types of cholesterol include high-density lipid cholesterol (HDL), which is known as good cholesterol, and low-density lipid cholesterol (LDL) or bad cholesterol. Cholesterol with an even lower density is known as very low-density lipid cholesterol (VLDL) and is also bad cholesterol (Table 3. 3).

The reason why HDL is called good cholesterol, is that it collects excess cholesterol from various organs and brings it back to the liver. On the other hand LDL and VLDL are called bad cholesterol because they take cholesterol from the liver to various organs. When bad cholesterol increases, excess cholesterol accumulates in various organs and blood vessels, which can lead to overweight and arteriosclerosis.

Although HDL is called good cholesterol, if it accumulates in excess amounts, this can cause a shortage in the amount of cholesterol in various organs leading to a weakening in the binding activity of the intercellular matrix and making the blood vessels susceptible to damage. Thus, it is necessary to keep a balance between the amounts of HDL and LDL. The desirable levels of HDL and LDL are 50 mg/dl and 100 mg/dl, respectively.

When high blood pressure continues for a prolonged time, the inner wall of the blood vessels is damaged and cholesterol enters the open wound. Cholesterol then begins to accumulate in the connective tissues and muscle cells, increasing in amount and forming aggregates. The wall of the blood vessels thickens, the vessel becomes narrow and the blood flow stops. This phenomenon is called thrombosis, which is a cause of cerebral infarction and ischemic heart disease.

3. 4. 4. Tea reduces bad cholesterol

Table 3.3. Content and types of good cholesterol (HDL) and bad cholesterols (LDL and VLDL) (Oguri, I. and Hara, Y.,1990)

Cholesterols	HDL cholesterol	LDL cholesterol	VLDL cholesterol	Chylomicron
Common name	Good cholesterol	Bad cholesterol	Bad cholesterol	
Density	High density	Low density	Super low density	Extra super low density
Protein content	55%	21%	8%	2%
Lipids contained	Phospholipid (30%)	Cholesterol (45%)	Fat (50%)	Fat (84%)
Synthesis organ	Liver, small intestine	Liver	Liver	Intestine
Functions	Recovery of choclesterl from tissues to liver	Transportation of cholesterol from liver to tissue	Transportation of lipids (inner lipids) synthesized in liver	Transportation of (external lipids) absorbed from foods
Effects on arteriosclerosis	Prevention, improvement	Promotion	Promotion	Promotion

It was found that green tea reduces bad cholesterol, which is a cause of arteriosclerosis. Rats were fed for 4 weeks on a high cholesterol diet containing 15% lard and 1% cholesterol with or without the addition of 1.0% crude catechins of green tea. Then the amount of total cholesterol, HDL and LDL, in the blood and livers of the rats was determined. Results are shown in Figure 3. 10. In rats fed the diet containing 1.0% catechins, the amount of total cholesterol, especially the level of LDL cholesterol, in the blood, decreased as compared with the control rats on the diet without green tea catechins (Figure 3. 11).

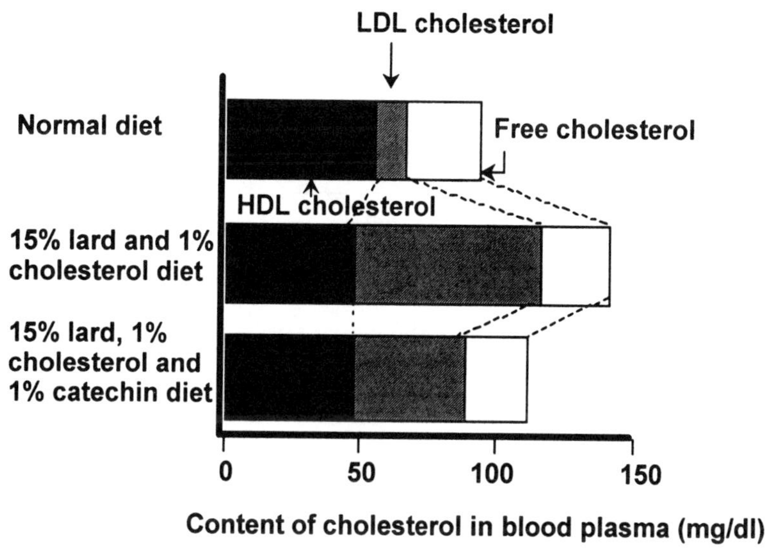

Figure 3. 10. Decrease of the LDL cholesterol level by tea catechins in blood of rats fed on high cholesterol diet. (Muramatsu, K., *et al.*, 1986).

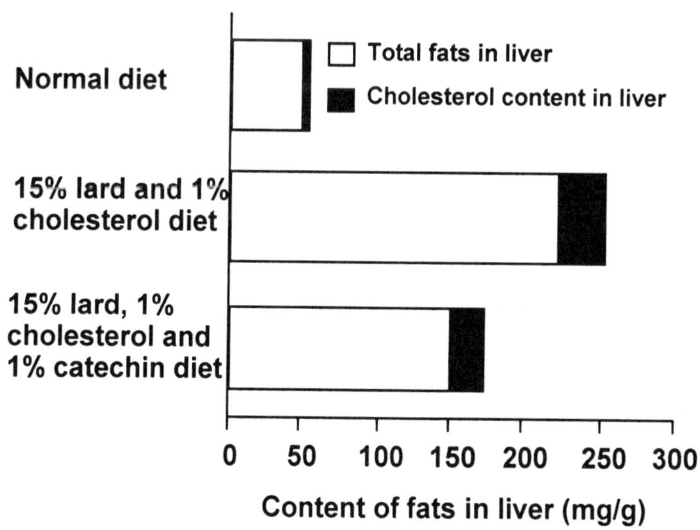

Figure 3. 11. Decrease of fat concentration by tea catechins in liver of rats fed

on a high fat/high cholesterol diet. (Muramatsu, K., *et al.*, 1986).

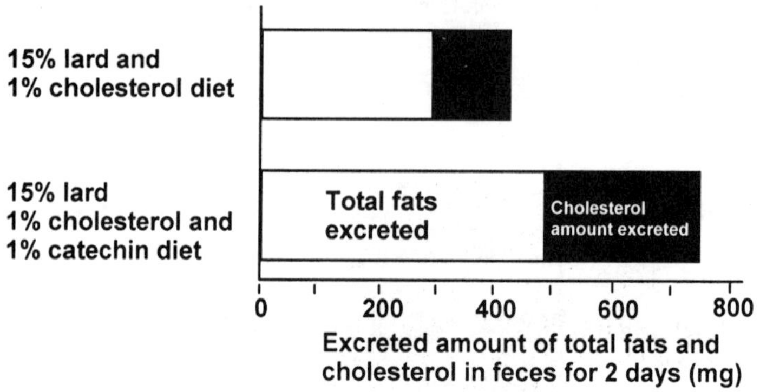

Figure 3. 12. Increase of total lipid and cholesterol levels by tea catechins in feces of rats fed on a high fat/high cholesterol diet. (Muramatsu, K., *et al.*, 1986).

Simultaneously, the amount of cholesterol in the liver decreased markedly. Also, the amount of cholesterol excreted in the feces, along with the amount of total lipids excreted, increased with the feeding of green tea, as shown in Figure 3. 12.

The above results indicate that the rise in LDL cholesterol in the blood and liver, induced by a high cholesterol diet, was suppressed by tea catechins due to the inhibition of absorption of cholesterol in the digestive organs. To determine what components in tea suppressed the rise in cholesterol, rats fed on a high cholesterol diet containing 0.5% or 1.0% EGCg were examined. The results showed that feeding of EGCg decreased the LDL cholesterol in the blood, while, on the other hand, the HDL cholesterol increased, as shown in Figure 3. 13.

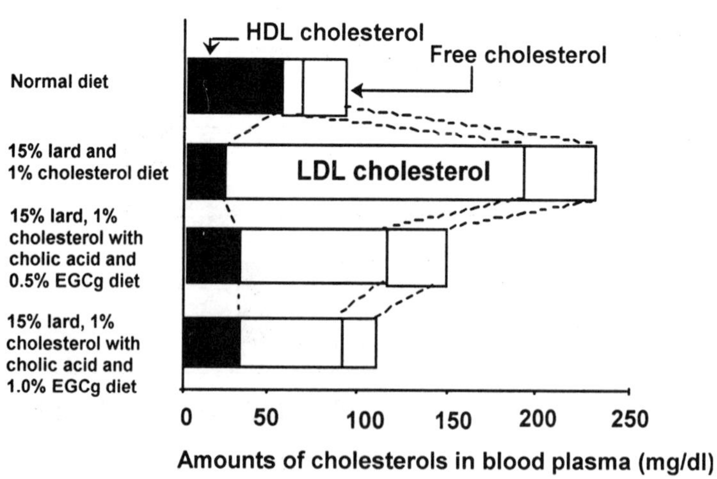

Figure 3. 13. Decrease of bad LDL cholesterol in blood plasma by tea catechin snd EGCg, in blood plasma

of rats fed on a high fat/high cholesterol diet. (Muramatsu, K., *et al.*, 1986).

3. 4. 5. Effect of tea on human blood cholesterol

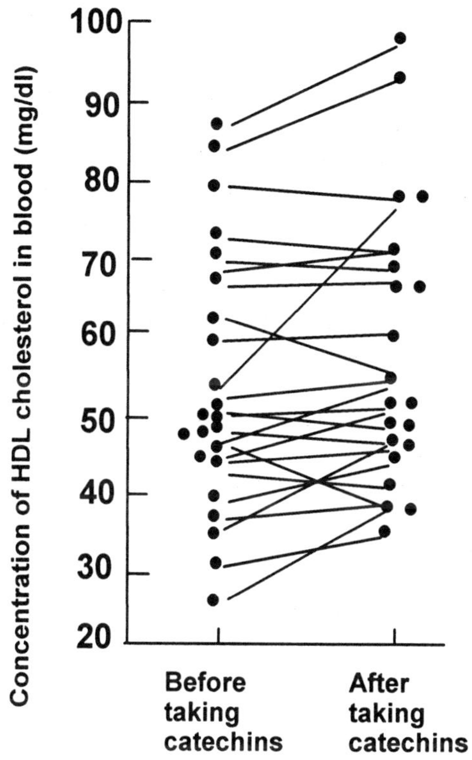

Figure 3. 14. Effect of tea catechins on increase of good HDL cholesterol in human blood. (Iwata, K., *et al.*, 1991).

Do tea catechins have a similar effect on human blood cholesterol? Thirty-three healthy adult persons (ages between 20 and 70) were given 400 mg catechins per day for 3 months, and the amounts of lipids in their blood was determined before and after taking tea catechins.

The results indicated that there was no change in the total amount of cholesterol in the blood before and after taking tea catechins. The amount of HDL cholesterol, however, increased slightly (Figure 3. 14), and the neutral fat decreased a little. This shows that tea catechin is also beneficially effective in the regulation of cholesterol in human blood.

Recently, it has been thought that the oxidation of LDL cholesterol may induce arteriosclerosis. Drinking of tea (green tea catechins, black tea polyphenols) was proved to delay the oxidation of LDL cholesterol.

These results confirm that tea catechins decrease the level of bad cholesterol and increase the level of good cholesterol, which leads to beneficial effects such as reduction of various diseases in the blood circulation system, particularly arteriosclerosis.

3. 5. PREVENTION OF SENILE DEMENTIA

3. 5. 1. Tea increases mental activity

In "Kissa-Yojoki" the author Yeisai claims that tea wards off sleepiness and raises the spirits. Most people find it difficult to fall asleep if they drink strong green tea before going to bed. Even when we are lying down, tea stimulates the general center in the brain, and we may start to think about the events of the day and what we will do the next day. It is the caffeine and related compounds contained in tea that stimulate the central nervous system in the brain and promotes mental activity.

Glucose present in the brain is metabolized to substances which become energy sources, and produce neural transmitters. Tea caffeine stimulates the production of these neural transmitters.

In Japan a TV program "Ultraeye" which investigated the effects of caffeine and green tea on the behavior of mice was aired by the national broadcasting system, NHK. In this experiment, mice were put in a maze and the time spent to arrive at the food goal through the maze spent was determined. With several repetitions of the trials, the time the mice took to reach the goal became shorter and shorter.

After these learning trials were repeated for a week, the time taken to reach the goal was one third or one fourth shorter than that of the initial trial. With mice drinking salt solution though, even after repeated trials for one week, the time taken to reach the goal hardly changed. However, when these mice were given green tea instead of salt solution, the time taken to reach the goal shortened. When mice were given caffeine, the time taken to reach the goal also shortened. These results indicate that tea catechins and caffeine can improve memory, judgment, thinking and the ability to learn from experience.

3. 5. 2. Alzheimer's disease

In Japan, where the already high number of elderly people in the population is still growing, senile dementia is a large-scale, serious problem. Institutions for elderly people, and some families caring for elderly relatives, have to cope with many problems that accompany senility. These include the charges forgetting the names of even their spouses and children and not recognizing who they are, demanding meals even after they have just eaten, accusing other family members of taking their purse which they themselves have misplaced, wandering around the neighborhood at all times of the day and night, and other such social problems.

If these symptoms are recognized in the initial stages and appropriate training or treatment is undertaken, in some cases it may be possible to return to normality. Recently, however, Altzheimer's disease has become a big problem in Japan. A characteristic of this disease is the appearance of many spots, known as senile spots (plaques) in the cerebral cortex of the brain, and the production of masses of abnormal fibers in nerve cells. The senile spots consist of amyloid protein surrounded by the projections of nerve cells, and these bring about the characteristic symptoms of senile dementia. Alzheimer's disease is thought to be hereditary and it is now believed that the gene responsible is on the human chromosome 21.

3. 5. 3. Tea prevents Alzheimer's disease

The nerve cells concerned with memory are located in the base part of the forebrain. It has been found that these nerve cells lose their functions in the brain of Alzhenimer's disease patients. Preventing the functions of these cells is thought to be a good way of treating Alzheimer's disease.

To achieve this purpose, a protein, the nerve growth factor (NGF) may be required. Recently teannin (Figure 3. 15), a component in tea (super green tea, Gyokuro) that is responsible for its sweet taste, was found to stimulate synthesis of this nerve growth factor.

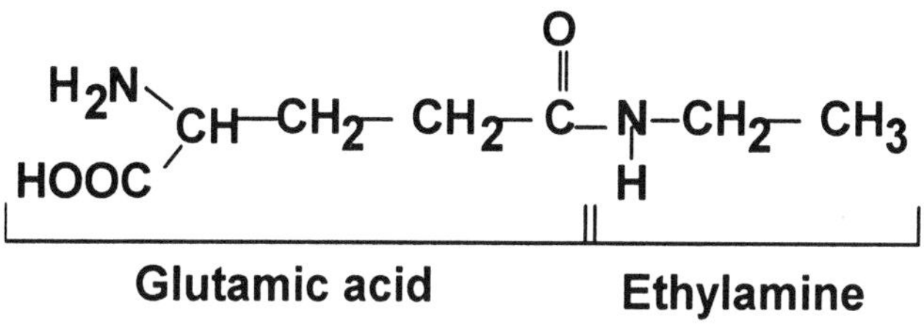

Figure 3. 15. Chemical structure of teanin (amino acid) contained in tea.

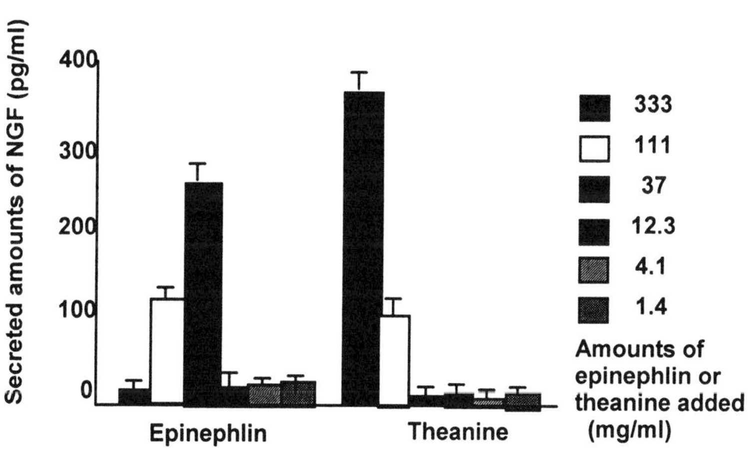

Figure 3. 16. Efect of teanin on increase of the secretion of neave growth factor (NGF) in astrocytes of rat brain, in comparison with the effect of epinephrine. (Kinoshita, T, *et al.*, 1994).

This fact was proven by an experiment in which the nerve cells, astrocytes, were removed from the brains of newborn rats, and cultured *in vitro* . When teannin was added to the culture medium, the synthesis of NGF was stimulated more strongly than when epinephrine was added (Figure 3. 16).

A research group in the Molecular and Cellular Biology Institute, University of Tokyo, cultured nerve cells taken from the hippocampus of the rat brain, and added to the medium amyloid protein obtained from senile spots of Alzheimer's disease. The projections of nerve cells were lost and the majority of the cells died. When the plant catechin, gallanan, produced in the Amazon district was added to the culture medium, the death of nerve cells was almost completely inhibited. These results suggest that tea catechins may be effective in protection against Alzheimer's disease.

When we drink tea, teannine is absorbed in the body, and transported to the brain where it stimulates astrocytes, thus increasing the secretion of nerve growth factor (NGF), enhansing the normal function of nerve cells, and resulting in protection from Alzheimer's disease.

APPENDIX 3

STORAGE OF TEA

During storage of tea for a long period, the fresh aroma of the green tea leaves gradually decreases, and the color and taste also deteriorate. The loss of quality of tea during storage occurs quite rapidly in green tea, and relatively slower in black tea. Oolong tea can be stored for a long time.

STORAGE AT ROOM TEMPERATURE

The quality of green tea changes considerably when stored at room temperature for several months. If, however, the water content in the tea is reduced to about 3% by drying, the deterioration may be reduced to some extent. Still, the loss of quality is much greater at room temperature than it is when the tea is stored at a low temperature.

STORAGE AT LOW TEMPERATURES

The deterioration of the quality of tea is due to the oxidation of components contained in tea. At low temperatures, the speed of the oxidation may be arrested to some extent. Tea can to be better preserved at lower temperatures. In the home, tea is kept in a refrigerator or a freezer to preserve quality. The temperature is generally 0°C to 5°C. When tea is stored together with other foods such as vegetables and fruits in the same refrigerator, the temperature in the refrigerator may rise and a lingering odor is transferred to the tea. In such cases, the tea should be put in a container.

HANDLING AFTER STORAGE AT LOW TEMPERATURES

If tea which has been stored under refrigeration is exposed to the outside air which is at a higher temperature, the vapor contained in the outside air produces dew on the surface of the cold tea and this is absorbed into the tea. To avoid this happening, it is better to put the tea in a cooler part of the room for about a day until the temperature of the tea is closer to that of the outside air.

4

INFECTIOUS DISEASES AND TEA

4. 1. PREVENTION OF INFLUENZA INFECTION

4. 1. 1. Preventive action of tea against influenza virus infection

Every year the coming of winter brings with it inevitable outbreaks of influenza throughout Japan. Different types of influenza virus are prevalent in different years; while one year type A may be difficult to prevalent, the next year type B is dominant and so on. Newspaper and television reports inform us of influenza epidemics that are particularly difficult to prevent in homes for the elderly people, many of whom have a weak constitution.

Various ways of preventing infection by the influenza virus are generally known, including wearing a mask and gargling. Green tea has been demonstrated to have a strong protective effect against the influenza virus. In this respect, green tea was found to be effective in human cells. When the influenza virus infects the cells, the spikes on the surface of the virus bind to the surfaces of the mucous membranes of human cells. Green tea catechins bind to spikes of the virus, which prevents the viruses from binding to human cells.

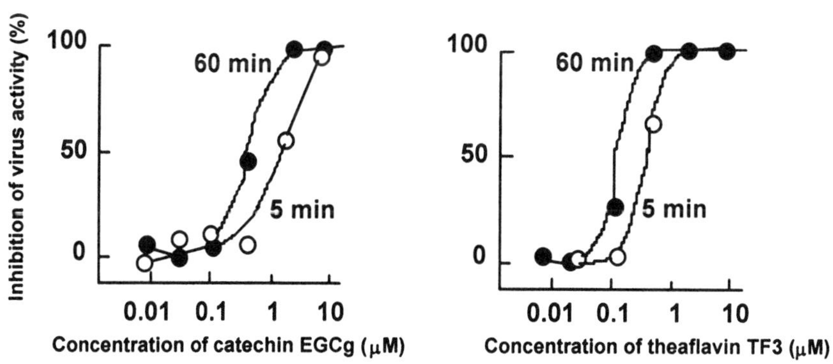

Figure 4. 1. Preventive effects of tea catechins, EGCg and TF3, on influenza virus infection. (Nakayama, M., *et al.*, 1993).

The catechin, EGCg, contained abundantly in green tea and the theaflavin TF3, contained in black tea, are effective in protecting against infection with the influenza virus. When catechins in black tea are added to canine kidney cells, MDCK, in culture, the cells do not become infected with influenza virus. EGCg in green tea and TF3 in black tea were found to be effective in the prevention of virus infection. As shown in Figure 4. 1, the effective concentrations of both polyphenols are as low as 0.5 to 1.0 mM.

4. 1. 2. Prevention of influenza virus in mice and pigs

Tea catechins showed protection against infection with influenza viruses in animal experiments. When mice were infected with influenza viruses, the death rate was 100%. However, when tea catechins were added to the viruses, the viruses did not infect the mice.

In pigs, it was also found that tea catechins inhibited the infection with influenza viruses. Pigs, in the same way as humans, are susceptible to frequent infection with the influenza virus. An experiment was carried out in which 1,200 pigs were divided into two groups, and over a period of 6 months pigs in one group were sprayed with a constant amount of catechin solution for 20 min, and every 30 min with a sprinkler, while those in the other group were sprinkled but not sprayed with catechin solution. It was found that in the catechin-sparayed pigs the antibody level for influenza virus in the porcine blood did not increase, and infection with the influenza virus was suppressed.

4. 1. 3. Tea prevents human influenza

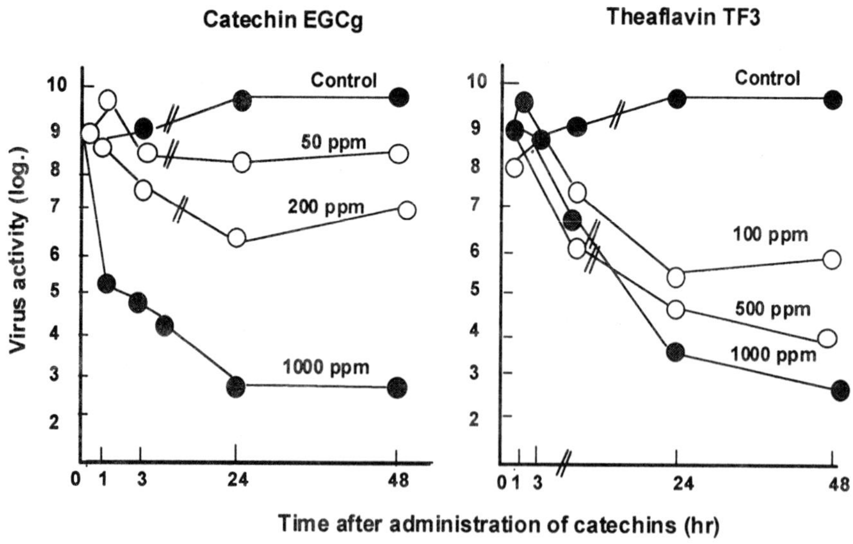

Figure 4. 2. Antibacterial effect of tea catechins, EGCg and TF3, on *Bordetela pertussis.*

(Horiuchi, Y., *et al.*, 1992).

In a clinical study with humans, gargling with black tea every day prevented infection with the influenza virus. The antibody level in the blood was found to be low. Gargling with all kinds of tea, such as green tea, black tea or oolong tea, if it is decocted, is effective. A concentration of about four times as strong as the usual drinking concentration is sufficient. It is important to use lukewarm tea and to put your head well back when gargling, to make sure that the tea reaches the depths of the throat..

In a certain elementary school in Shizuoka Prefecture, gargling was carried out by all school children twice a day, once when they arrived at school and then again before they returned home. Even during times when influenza infention was epidemic and many schools or classes had to close, the school where the children were gargling did not have to close. Gargling with green tea not only prevented infection with influenza viruses, but also prevented progression to a severe state as well.

Tea catechins ,in particular,EGCg in green tea and TF3 in black tea (Figure 4. 2), are also markedly effective in killing the bacteria which cause the infectious childhood disease whooping cough).

4. 2. PREVENTION OF DENTAL CARIES AND BAD BREATH

4. 2. 1. Prevention of dental caries in school children

It was found that tea is effective in the prevention of dental caries. Eating too many sweet foods can lead to dental caries if proper care is not taken to prevent them by brushing and cleaning the teeth. Dental caries are formed when bacteria produce a water-insoluble glucan from sugars in foods (Figure 4. 3). This glucan is called dental plaque, and it provides an environment in which the bacteria proliferate and produce acids, resulting in the enamel layer on the tooth dissolving and formation of cavities.

To prevent dental caries, it is necessary to remove the bacteria present in the mouth, not to consume too much sugar, and to stop the production of dental plaque. Drugs have been developed which counter these factors that produce dental caries. Sweeteners that lower the sugar content and substances which coat the surfaces of the teeth to prevent the attachment of bacteria are on the market.

The suppressive effect of tea drinking against dental caries was investigated in a primary school in Chiken in Kagoshima Prefecture. School children in the first grade were divided into two groups. In one group the children drank a cup of tea at lunchtime and in the other the children did not drink tea. After one year, a dentist examined the number of dental caries in all children in both groups. The number of dental caries in the children drinking tea was about half (50%) that of the children who were not drinking tea.

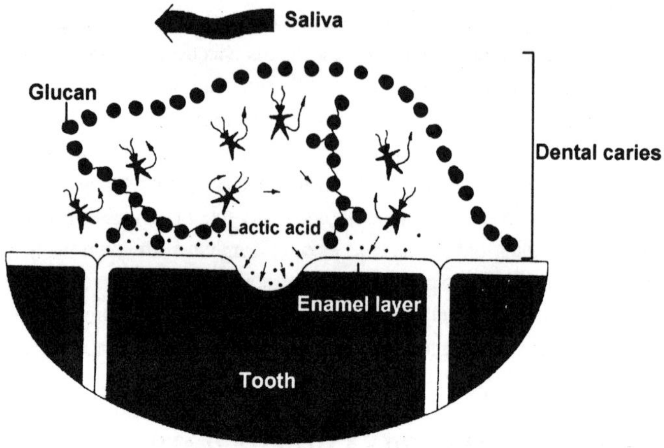

Figure 4. 3. Diagram showing the development process of dental caries by bacteria. (Oguni, I. and Hara, Y., 1991).

Table 4.1. Effect of tea catechins (10 mM) on the activity of enzyme produced by bacteria in the formation of dental caries

Tea catechins	Inhibitory rate (%)
(-)-Epicatechin (EC)	42.3
(-)-Epigallo catechin (EGC)	24.7
(-)-Epicatechin gallate (ECg)	83.0
(-)-Epigallocatechin gallate (EGCg)	75.0
Theaflavin (TF)	98.3
Theaflavin gallate A (TF2A)	97.3
Theaflavin gallate B (TF2B)	97.8
Theaflavin digallate (TF3)	98.2

A study to examine the effects of tea on the prevention of dental caries in infants was carried out in Haibara Village in Shizuoka Prefecture for a period of five months, beginning in September 1973.

Sixty-one infants born during this period were given 100 ml of tea made from the fourth infusion of low-caffeine tea leaves, every day, and the frequency of dental caries was examined. The results showed that the frequency of the infants with dental caries was 0% (control: 3 to 6.5%) at 1.6 years old, 7.7% (control: 21 to 29%) at 2 years old, and 21.9% (control: 55 to 66%) at 3 years old. The average number of teeth with caries per baby also decreased to 0.13 (control: 0.7) at 2 years old and 0.65 (control: 3.3) at 3 years old. These inhibitory effects on the development of dental caries may be due not only to catechins in tea but also to fluorine in the tea water.

4. 2. 2. Inhibition by tea of the formation of dental caries

The bactericidal effect of tea catechins against dental caries was confirmed by the following experiment. A thin glass plate was inserted in a sugar-containing medium and bacteria were grown on the plate. After incubation for three days, the formation of glucan masses produced by the bacteria was examined. When tea was added to the medium, at the usual drinking concentration, the formation of glucan was completely inhibited. Even when the concentrations of tea were diluted twice, the inhibiting effect of tea did not change. No significant differences in effectiveness were found among green tea, black tea, oolong tea and Pu-er tea.

The inhibiting effect of tea catechins against the formation of dental caries by bacteria may be as a result of their direct killing effect and also their suppressive effects on the activity of enzymes involved in producing dental caries. The bactericidal effects of four kinds of tea at the usual drinking concentrations were examined. As shown in Fig. 4. 4, green tea was found to have the strongest bactericidal effect, followed by Pu-er tea and black tea. Oolong tea had a week bactericidal effect, and probably suppressed the formation of dental caries by suppressing the enzyme activity involved.

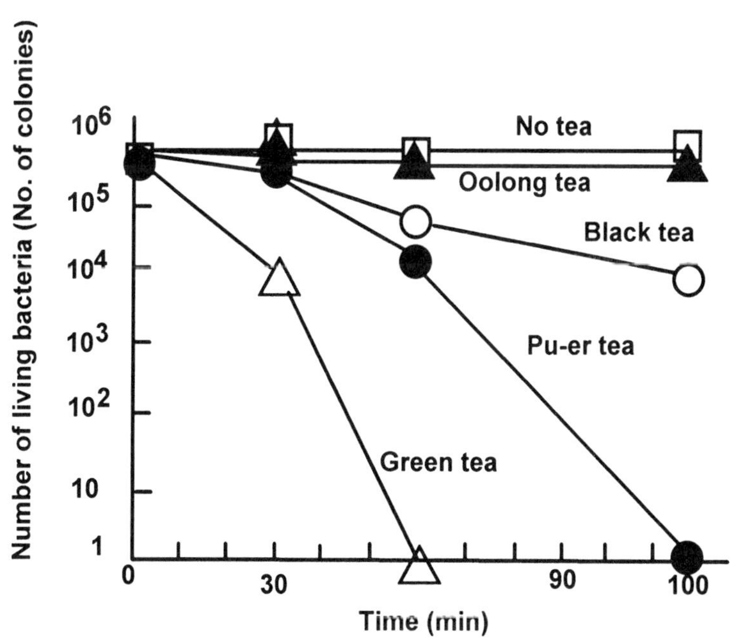

Figure 4. 4 Antibacterial effect of various teas on dental bacteria (*Streptococcus mutants*). (Ishigami , T. and Hara, Y., 1993).

In addition to the direct bactericidal effects of tea catechins on the formation of dental caries, tea catechins suppressed the activity of an enzyme produced by bacteria and involved in the formation of dental caries by sugars. This experiment was carried out by using sugar with isotopic carbon and by determining the radioactivity of the dental caries formed. The results are shown in Table 4. 1. The formation of dental caries decreased with the addition of tea catechins and theaflavins at a concentration of 10 mM.

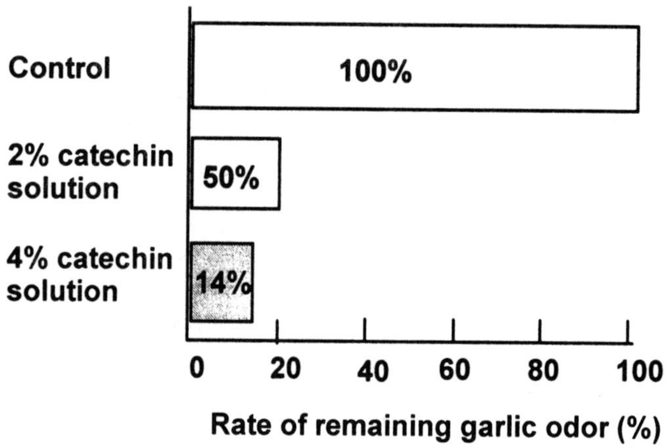

Figure 4. 5. Deodonizing effect of tea catechins on garlic odor. (Kawakami, M.,*et al.*, Unpublished)

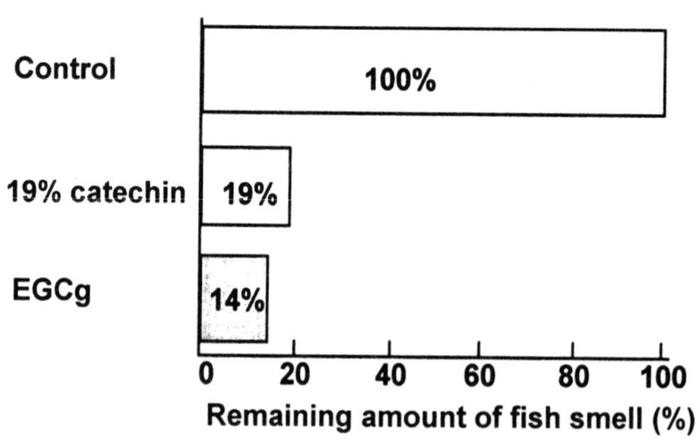

Figure 4. 6. Deodorizing effect of tea catechins on fish odor. (Kawakami, M., *et al.*, unpublished).

Table 4.2. Comparison of minimum growth inhibiting concentrations of tea catechins against food poisoning bacteria (Hara, Y. and Ishigami, T., 1989)

Food poisoning bacteria	Minimum growth inhibitory concentration (ppm)			
	EC	ECg	EGC	EGCg
Staphylococus aureus IAM 1011	>800	800	150	250
Vibrio fluvialis JCM 3752	800	300	300	200
Vibrio parahaemolyticus IFO 12711	800	500	300	200
Vibrio metschnikovii IAM 1039	>1,000	>1,000	500	1000
Clostridium perfringens JCM 3816	>1,000	400	1000	300
Bacillus cereus JCM 2152	>1,000	600	>1,000	600
Plesiomonas shigelloides IID No. 3	700	100	200	100
Aeromonas sobria JCM 2139	>1,000	700	400	300
Clostridium botulinum A, B mix.	>1,000	<100	300	<100

When catechins or EGCg, were added to a solution with a fish odor, the fish odor diminished (Figure 4. 6). With the addition of a 0.1% catechin solution, the fish odor decreased to 19%, that is one fifth of the original; and with the addition of EGCg, the fish odor decreased to14% of the control value.

Another dental hygiene problem, in addition to dental caries, is bad breath. In particular, garlic and fish odors can be offensive to some people in certain circumstances. When catechins were added to garlic solutions at concentrations of 2% or 4%, the garlic odor decreased (Figure 4. 5). With the addition of 2% catechins, the garlic odor decreased by half. The addition of 4% catechins decreased the garlic odor to 14% of the control value.

4. 3. PREVENTION OF FOOD POISONING

4. 3. 1. Foodborne pathogenic bacteria

The bacteriocidal activity of tea has been described in the section on dental caries. Tea also has a strong bactericidal activity against a wide range of food poisoning bacteria. As shown in table 4.2, tea catechins, EC, ECg, EGC and EGCg showed bactericidal activity against many food poisoning bacteria such as *Staphylococcus aureus, Clostridium perfringens, Bacillus cereus, Vibrio parahaemolytics* and *C. botulinum.*

When we consume foods that are contaminated with these bacteria we are affected by food poisoning.

4. 3. 2. Bacteriocidal activity of tea

After a food poisoning incident in Japan in 1984, which was caused by *C. botulinum* in a lotus root product, it became compulsory to sterilize all neutral (low acidic) drinks (including tea drinks) at 120°C for more than four minutes. *C. botulinum* is resistant to heat and when the bacteria are germinating/growing they produce a strong poison than can be deadly to humans.

As shown in Figure 4. 7, when *C. botulinum* was added to various tea drinks such as black tea with low sugar, sweetened black tea, green tea, or oolong tea, and left to stand under constant conditions, without heating, the (non-growing) spores of *C. botulinum* were gradually killed by tea catechins. When the tea drinks were divided into a catechin fraction and a non-catechin fraction, then *C. botulinum* was added to each fraction and left to stand, it was found that the catechin-containing fraction had the bactericidal activity.

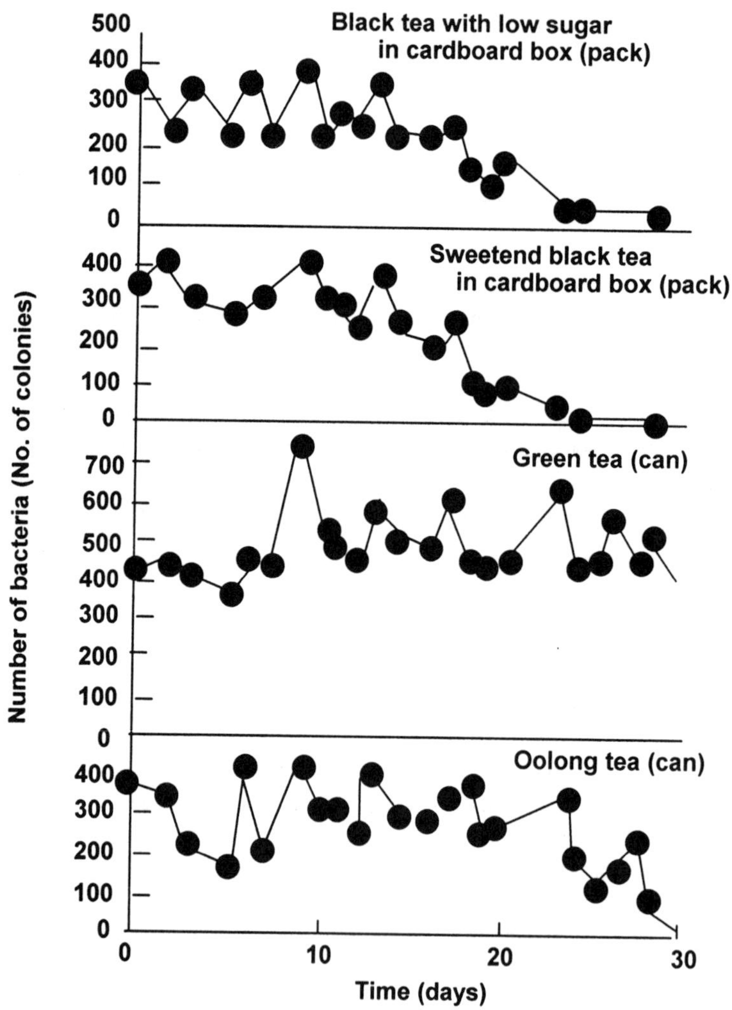

Figure 4. 7. Gradual decrease in spores of *Clostridium botulinum* added to various tea drinks (Hara, Y. *et al.* 1989).

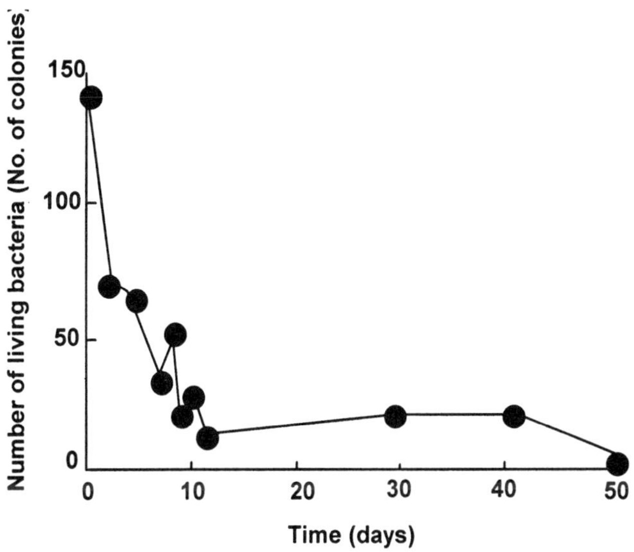

Figure 4. 8. Graph showing the decrease in the number of *C. botulinum* added to green tea drink and treated at 85 ℃ for 30 min. (Hara, *et al.*, 1989).

When green tea catechins were added to *C. botulinum*, and the mixture left to stand without heating, the bacteria were not completely killed after 30 days, but when this was repeated with heating at 85°C for 30 min, the bacteria were completely killed (Figure 4. 8). The explanation for this is that heating causes the non-growing bacterial spores to germinate (begin to grow), and they are killed when germinate.

4. 3. 3. Tea kills bad bacteria and promotes good bacteria

The food poisoning bacteria are called "bad" bacteria. Tea has a strong ability to destroy these bad bacteria. The human intestine harbors, in addition of these bad bacteria, some good bacteria. Tea has no bactericidal activity against these good bacteria (Table 4. 3). The reasons why tea catechins have differential activity against bad and good bacteria are as follows: the bad bacteria produce ammonium and sulfides that create an alkaline state in the intestine. On the other hand, the good bacteria produce acidic substances. Tea catechins are active under alkaline conditions and not active under acidic conditions.

Table 4.3. Minimum growth inhibitory concentration of tea catechins for harmfuland beneficial bacteria in the intestine (Hara, Y. and Ishigami, T., 1989)

Bacteria	Minimum growth inhibitory concentration (ppm)
Harmful bacteria	
Staphylococcus IAM 1011	450
Vibrio JCM3752	200
Vibrio IFO12711	200
Vibrio IAM 1039	500
Clostridium JCM 3816	400
Clostridium A, B mix	>100
Bacillus JCM 2152	600
Plesiomonas IID No.3.	100
Aeromonas JCM 2159	400
Beneficial bacteria	
Lactobacillus JCM 1102	>1,000
Lactobacillus JCM 1059	>1,000
Lactobacillus JCM 1183	>1,000
Bifidobacterium JCM 1255	>1,000
Bifidobacterium JCM 1275	>1,000
Bifidobacterium JCM 1217	>1,000

4. 4. IMPROVEMENT OF INTRAINTESTINAL BACTERIA AND PREVENTION OF ODOR

More than one hundred kinds of bacteria are living in the human intestine. Numbering as many as 10^{14}, they are closely related to human health and various lifestyle-related diseases, aging and immune reactions. Many bacteria are detected in the feces of any healthy adult person. The most predominant bacteria include *Bacteroidaceae, Eubacteria, Peptococaceae, Clostridia* and *Bifidobacteria.*

Changes occur in the type and number of these bacteria in the intestine with age. Good bacteria such as *Bifidobacteria* decrease gradually. On the contrary, putrefactive bacteria such as *Enterobacteria, Bacteroidaceae, Clostridia,* and *Enterobacteriaceae* increase. Tea catechins have a strong antibacterial activity against food poisoning bacteria and putrefactive bacteria and no effect against lactic acid bacilli such as *Bifidobacteria and Lactobacilli* The effects of tea catechins on the intra-intestinal bacteria and their preventive action against bacterial putrefaction were examined in elderly people who were taking tea catechins.

4. 4. 1. Effect of tea catechins on intrainestinal bacteria

Koichi Goto in the Hamamatsu-Juji-no-En Clinic, with the collaboration of Seirei-Mikatahara Hospital, has examined the effects of tea catechins on intraintestinal bacteria. Elderly bed-ridden patients took a 100 mg capsule of tea catechins (Polyphenon 60) three times a day at meal times (breakfast, lunch and supper) and changes in their intraintestinal bacteria were examined. Their total daily intake of tea catechins was 300 mg each day, which corresponds to five to six cups of tea. All subjects were persons aged from 51 to 93 years old (average 70.5 years old), 10 females and 5 males, who became bed-ridden due to cerebral vessel injury or other diseases. They were given 1,000 k cal of liquid food every day by way of a nose tube or a stomach tube. This was supplemented with 30 ml of soybean sauce and 1,300 to 1,600 ml of water. Under these constant controlled conditions, the feces were investigated every day to determine the kinds and numbers of bacteria present.

Results are shown in Fig. 4. 9. During the three weeks after tea catechins were given, good bacteria such as lactobacilli and *Bifidobacteria* increased gradually, while bad bacteria such as *Clostridia and Escherichia coli* decreased. It was proved that tea catchins had anti-bacterial effects only on bad intra-intestinal bacteria, and had beneficial effects on good bacteria.

With these changes in intra-intestinal bacteria, the amounts of organic acids in feces increase and become acidic with a resulting decrease in odorous constituents such as ammonia and sulfides. There was also a reduction in the amounts of indol, sakatol and phenols which cause the putrefactive odor of feces (Figure 4. 10).

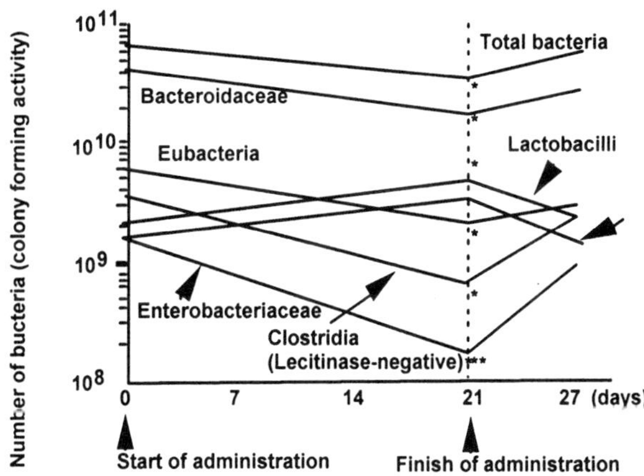

Figure 4. 9. Change in the number of intraintestinal bacteria in the feces of elderly bed-ridden patients taking 300 mg tea catechins a day. (Hara,, Y., 1999).

Nursing staff gave good evaluations on the effects of taking tea catechins, reporting that there was no fecal odor, and the amount of feces increased.

When tea catechin administration was stopped, examination of intra-intestinal bacteria indicated a tendency to return to the original state after one week. On the basis of these results, the Hamamatsu-Juji-no-En Clinic decided to give tea catechins to patients and food servers and obtained good results in subsequent feces examinations.

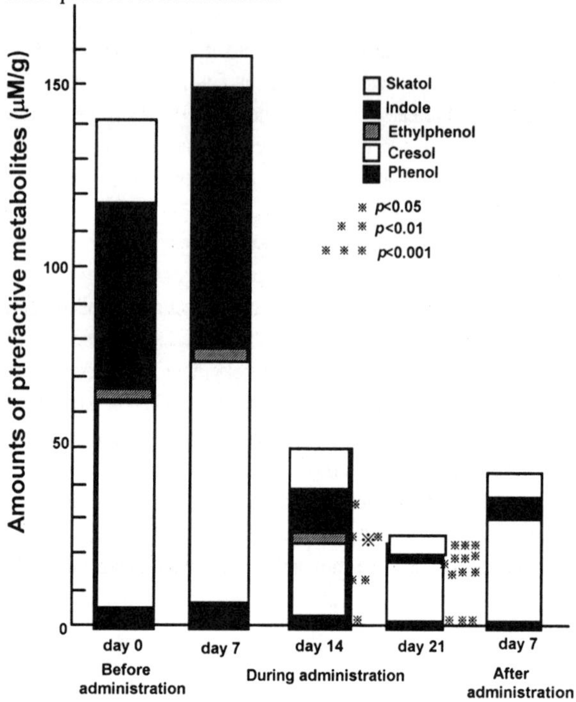

Figure 4. 10. Decrease in putrefactive metabolites in the feces of elderly bed-ridden patients taking 300 mg tea catechins a day. (Hara, Y.,1999).

Further excellent results on the activity of tea catechins in controlling intestinal conditions and evacuation were obtained when 500 mg capsules of tea catechins were administered to 37 healthy adults for 3 months.

Even those who were having trouble with constipation, going for three to four days between bowel movements, came to have a motion everyday after taking catechin capsules for a few days. Drinks containing lactic bacteria, and foods such as yogurt, oligosugars and fibers are recommended for controlling the state of the digestive tract. A similar effect may be obtained by taking tea catechin capsules.

4. 5. PREVENTION OF INTRA-HOSPITAL INFECTIONS

4. 5. 1. Intra-hospital infective bacteria, MRSA

Since 1980 patients infected during their hospital stay with methicillin-resistant *Staphylococcus aureus* (MRSA) have increased. When patients who have reduced immune system activity become infected with MRSA immediately after an operation, the condition can become serious. Procedures to prevent infection with MRSA are required. MRSA bacteria were plated on agar medium containing various concentrations of Polyphenon 100 with more than 90% tea catechins.. Various antibiotics were added to these plates and their inhibition activity was determined.

MRSA is a very common type of bacterium present everywhere. This bacterium resides in the affected purulent area of inflammation, on the skin, and in the nose and pharynx. The bacterium has highly pathogenic properties and is the cause of various diseases such as infectious diseases of the skin and flesh, disseases of the intestinal tract (food poisoning), septicemia, endocarditis, cerebromeningitis, and infectious diseases of the respiratory organs and those of the urinary passages.

As long as the patient is resistant enough, there will not be any proliferation of the bacteria nor any occurrence of infectious diseases. However, infections with MRSA expand rapidly in hospitals where patients are treated with antibiotics to kill the infectious bacteria. Use of antibiotics in this way, though, may give rise to antibiotic-resistant bacteria during the treatment period.

4. 5. 2. Tea kills MRSA

It was found that tea catechins, EGCg in green tea and TF3 in black tea, had a strong germicidal action against MRSA. The effective amount of tea catechins for this purpose is one cup of green tea or 2 cups of black tea at the normal drinking concentration. Tea catechins were effective against MRSA, not only on their own, but they were even more strongly effective when combined together with oxacillin, methicillin or aminobenzyl penicillin (Figure 4.11).

In the Fukuoka University Hospital, 23 patients (almost all more than 70 years old), in whom MRSA had been detected in their nasal cavities or trachea, but in whom no symptoms of MRSA disease had appeared, were given a solution containing tea catechins (polyphenon 100) to inhale. The patients inhaled two or three times a day, indicating the consumption of a total daily amount of 20 to 40 mg catechins, until finally MRSA was not detected. Under these conditions, when inhalation of catechins was stopped, MRSA infection re-occurred but when the dose was increased to 80 mg a day, MRSA disappeared by the 6th day, and there was no re-occurrence. Subjects were free from MRSA infection for 10 days after stopping inhalation.

4. 6. Prevention of allergy

4. 6. 1. Mechanisms of allergic reactions

Every year in the early spring when fresh green growth is abundant an increasing number of people in Japan are plagued with pollen allergies. Allergy symptoms occur due to reactions to the pollens of Japan cedars and other trees, dust, and mites. When these allergens enter the mouth and nose, a specific antibody, IgE, is produced, which binds to the surface of mast cells present in the mucous membranes of the throat and the dermis under the skin.

If these same allergens enter the body, the IgE bound on the surface of the mast cells reacts with them and the mast cells excrete histamine and related substances that cause inflammation and the swelling of blood vessels. As a result, the reverse side of the eyelid and the mucous (conjunctival membrane) covering the white of the eye are congested, and this leads to tears, sneezes, snivels and itches. Pollen allergies are a result of the reactions in human body simultaneously producing allergic rhinitis and allergic conjunctivitis.

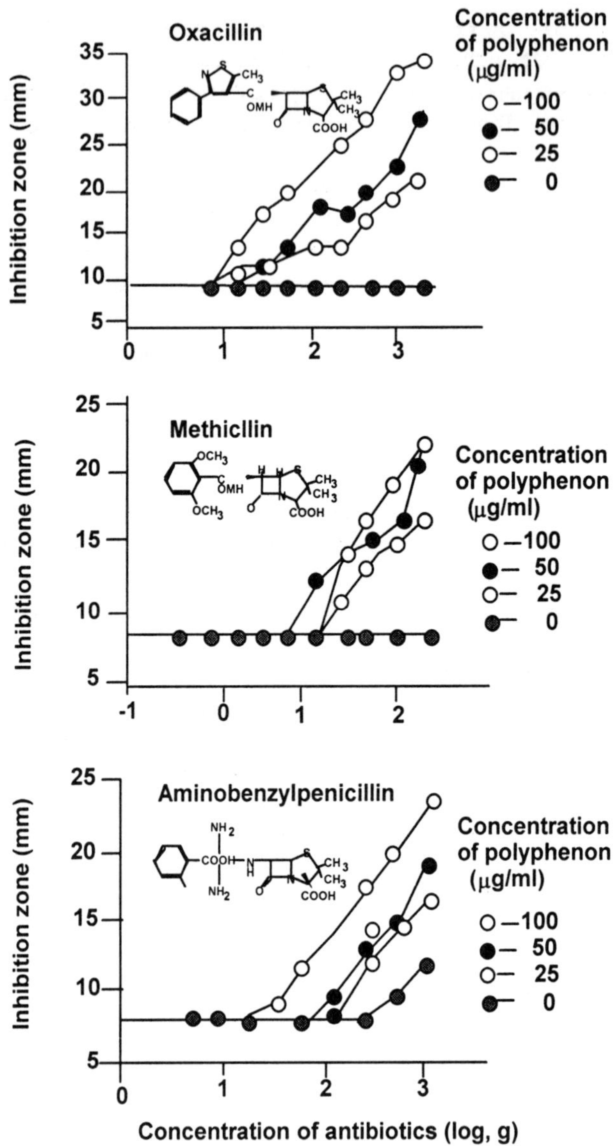

Figure 4. 11 Synergistic effects of various antibiotics and tea catechins (Polyphenon 100) on MRSA (methyciline resistant *Staphylococcus aureus*). (Shimamura, T. and Hara, Y.,1996).

Food allergies induced by eggs, fish, meat, soybeans, and other proteins, and inflammations induced by cosmetics and metal accessories, and rashes which occur when certain materials come in contact with the skin, are all produced by similar reactions related to the immune system.

4. 6. 2. Oolong tea prevents allergic reactions

Some investigators working in the Plant and Tea Experimental Stations in the Ministry of Agriculture, Forestry and Fisheries, and at the University of Shizuoka, have isolated from experimental mice, the mast cells which play the main role in the allergic reaction. They have cultured these cells to induce the secretion of histamine by reacting them with allergic antigens or IgE antigen. When a hot-water extract of tea was added to the cells, the amount of histamine produced was suppressed.

On comparing the suppressing action on histamine production by various kinds of tea, it was found that oolong teas produced in Taiwan (Tungting oolong and Chin-Hsin oolong) strongly suppressed histamine production. A popular kind of green tea, Yabukita, had no effect. They found that the effective substance in Taiwan oolong teas was epigallocatechin-4-*o*-methyl gallate.

However, whether or not the catechins present in oolong teas, which are different from those present in green teas, and are effective in the suppression of histamine production in mast cells, will also be effective in the prevention of pollen allergies remains to be seen.

Similar experiments were carried out in the Health and Environmental Center of Shizuoka Prefecture using experimental rats. They found strong suppressive effects on histamine production in mast cells with extracts from all eight kinds of green teas tested. Sencha (non fermented tea), Hojicha (roasted tea), Fukamushi-cha (heavily steamed tea), Kamairi-cha (pan fired tea), and semi-fermented tea extracts were active. In their study, Sencha, non fermented green tea, was the most effective in suppressing histamine. It is interesting that tea is effective in the prevention of allergic diseases such as pollen allergy.

4. 7. PREVENTION OF INFECTION WITH HIV

4. 7. 1. HIV

In recent years, the numbers of patients infected with human immunodeficiency virus (HIV) have increased throughout the world. The HIV virus originally proliferated in monkeys in Africa before the infection was transferred to humans. It spread widely, initially in America, Europe and East-Asian countries, and later throughout the world. Infection was usually by way of sexual intercourse or transfusions with infected blood. In Japan there was also a case of hemophilia patients being infected with HIV by transfusions with blood contaminated with HIV. The persons in change of the heat-inactivation of viruses in commercial blood samples were brought to court, and the chiefs of sections in charge in the Ministry of Health and Welfare and the chiefs in the HIV Investigation Committee were seriously criticized by the mass media for their response to this problem.

The HIV virus belongs to the retrovirus group and its genetic material is RNA instead of the DNA contained in most organisms. In the proliferation of this virus, RNA in the virus is reversely transcribed to form DNA. This reaction is called reverse transcription, since the reaction is a reverse reaction to the usual one from DNA to form RNA. The DNA reversely transcribed in HIV usually transcribes to form RNA, which then translates to form proteins in the ordinary way.

4. 7. 2. Tea inhibits the growth of HIV

The HIV virus has an enzyme, reverse-transcriptase, which carries out the reaction of transcribing RNA to reverse-transcript DNA. If this enzyme activity can be inhibited, the growth of HIV may be inhibited without any influence on normal tissues or cells. The well-known HIV treatment drug, azidothymidine (AZT) is an example of drugs which inhibit reverse-transcriptase, which thus inhibit strongly the growth of HIV. However, although these drugs strongly inhibit proliferation of the virus, they have a variety of side effects on the hematopoietic and other normal systems.

Tea catechins inhibit HIV growth with neither cellular toxicity nor side effects on other physiological functions of patients. An experiment was carried out by researchers in the Aichi Cancer Center Institute. They found that the addition of tea catechins, ECg or EGCg, inhibited the reverse-transcriptase activity of HIV *in vitro* (Figure 4.1 2).

EC had no effect in this experiment. The mode of infection and growth, and the antigenicity of HIV change easily. The length of time from the infection with HIV to the appearance of the disease is very long. Thus there are still many unknown factors related to prevention of infection, and medical treatment of HIV which remain to be resolved. Even against this background it is still valuable to emphasize that tea catechins are effective in the suppression of HIV growth during its early stage.

4. 8. PREVENTION OF FORMATION OF WARTS AND INFLAMMATION

4. 8. 1. Warts and inflammation

Warts and inflammation are produced by the growth of viruses and bacteria which infect skin tissues and spread in a red or red-brown rash on the skin surface, accompanied by itchiness or pain. These are different from the allergic skin inflammations induced by the production of allergens in mast cells in the immune system of the body against the antigens of external pollens and food factors. In the case of warts and inflammation, it is necessary to kill the viruses or bacteria which are the cause of the condition.

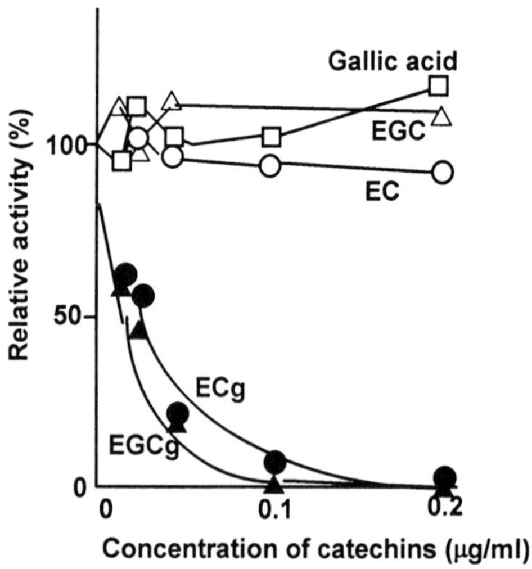

Figure 4. 12. Effect of tea catechins on suppression of reverse transcriptase activity of HIV (human immunodeficiency virus). (Nakane, H., *et al.*, 1994).

Usual medical treatments include ointments that are directly applied to the skin surface; alternatively oral medicines containing antibiotics or adrenal cortex hormone may be used. However, these drugs and medicines may produce an allergic reaction making the symptoms worse.

4. 8. 2. Condyloma in human genital organs

Condyloma inflammations occur in the external sexual organs, the penis in males, the vagina and the labium in female, and in the area surrounding the anus. The inflammation is caused by infection with papilloma virus (HPV) and appears 1 to 6 months (average: 3 months) after virus infection. No subjective symptoms are detected at first. In the infected area, nipple-like or cockscomb-like warts form and the area become red and inflamed. To determine whether the inflammation is caused by HPV or not, DNA of HPV obtained from an inflamed part of the tissue was examined.

It was found that a tea catechin mixture, Polyphenon E (Mitsui Norin, Fujieda, Japan) which contains four kinds of polyphenols, was effective in the treatment of condyloma. EC, ECg, EGC and EGCg from green tea can be safely used, without side-effects, in the form of a cream, jelly or milky ointment or as a capsulated suppository applied to the inflamed area by the patients themselves.

The safety of Polyphenon E was examined with mice. A group of 50 mice had a tea catechin ointment applied continuously for 7 days, and the cervical area of the uterus was examined to confirm the safety of Polyphenon E.

4. 8. 3. Prevention of human condyloma

The effects of the tea catechin ointment and suppositories on condyloma in human vulva, vagina and uterine cervical tissue were clinically examined at the Peking Cancer Center. Results are shown in Table 4. 4

The catechin cream was applied to 80 female patients; the vulva of 51 patients, vagina of 22 patients, uterine cervical tissue of 7 patients, once a day continuously for 2 months. Forty (50%) of patients were completely healed twenty-three (33%) were healed in more than 50% of the inflamed area, and seventeen patients (24%) showed no change.

Table 4.4. Effect of treatment with tea catechins on human external sexual organs (Tei, S. and Hara, Y. : original table)

Exp. No.	No. of inflamed parts	Completely healed	50% healed	No effect	Effective rate (%)
1	17	5	4	8	78
2	41	25	11	5	92
3	22	10	8	4	81
Total	80	40	23	17	86

The effects of tea catechins on other warts and inflammations have not yet been investigated, but if tea ointment and cream can be used easily by anyone to suppress inflammation, pain and itchiness, without side-effects, this is a very positive thing.

APPENDIX 4

PROVERBS ABOUT TEA

From ancient times in Japan, there have been many proverbs concerned with tea. Some of them are as follows:

TEA RICE AT BREAKFAST

Tea rice is a very easy meal to prepare for breakfast. The proverb means that things are very easy and there are no problems.

TEA RICE IN KYOTO

People living in Kyoto sometimes say to their guest "Please have some tea rice" although they do not want to be hospitable. The proverb means that they wish to be amiable but they have an empty heart.

STRONG TEA IS POISON FOR THE EYE, BUT MEDICINE FOR THE SPIRIT

Strong tea wakes you up and may cause insomnia, but it refreshes the spirit.

ALCOHOL SHOULD BE BOUGHT IN LIQUOR STORE AND TEA SHOULD BE BOUGHT IN A TEA SHOP

It is best to buy things from specialist shops where you can get comprehensive advice on the products.

USING A TEA BAG FOR THE HEAVY STONE WHEN MAKING PICKLED RADISH

In the process of making pickled radish a heavy stone is used to press down the radish in a pail. A tea bag is too light to use as a press weight, so the proverb means something which is good-for-nothing, is not effective or does not work.

USED TEA LEAVES CAN BE USED AS FERTILIZER

Even used tea leaves can be put to good use. We should look closely at things we throw away to see if there is some way we can use them again.

A TEA STALK FLOATING VERTICALLY IN YOUR TEA IS A GOOD OMEN

When you prepare tea by pouring hot water onto tea leaves in a cup, if a tea stem stands vertically in the hot water and you do not tell anybody, you may have good luck.

5

ANTIBACTERIAL AND ANTIOXIDATIVE ACTIVITY OF TEA

5.1. ANTIOXIDATIVE ACTION OF TEA

5.1.1. Active oxygen in the body

In the human body the energy necessary for living is produced by the burning of nutrients in the body using oxygen from the air inhaled through respiration. Some of the oxygens taken into the body, however, change as a result of metabolic processes or organisms exposed to natural radiation or UV-irradiation to superoxide which is a dangerous substance. The superoxide thus produced destroys cells in the body, inactivates many enzymes and alters the DNA and RNA of genes. When superoxides combine with lipids in the body superoxide lipids are produced, and these may initiate carcinomas and genetic diseases, weaken immunological activity, lead to various diseases related to aging, and shorten the lifespan.

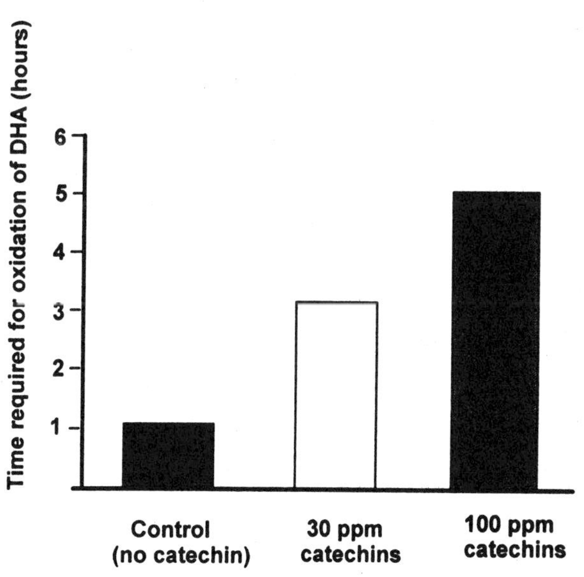

Figure 5. 1. Effect of tea catechins on vitamin E in blood plasma and red blood cells in rats fed on high fat diet. (Nanjo, F., *et al.*, 1993).

In the human body, several enzymes are produced that prevent the formation of and suppress the activity of superoxide. However, the activity of such enzymes gradually diminishes with age, resulting in progression of senescence, carcinogenesis and other age-related diseases. To supplement the activity of these enzymes, it is important to take vitamins A (carotene), C, and E. These vitamins have a strong anti-oxidative activity and prevent the superoxidation of lipids. Tea contains an abundant amount of these vitamins and also the catechins described above, making it a strong antioxidant.

5. 1. 2. Antioxidant action of tea

A diet rich in fatty acids, containing 30% palm oil or 30% perilla oil, was fed to rats. Tea catechins were added to this diet and the amounts of vitamin E in plasma and red blood cells were determined. When perilla oil, which contains many unsaturated fatty acids, was given to rats, vitamin E in the plasma and red blood cells was more oxidized and depleted than when palm oil, containing many saturated fatty acids was added (Figure 5. 1).

When tea catechins were fed in the diet of these rats, the reduction of vitamin E in the plasma and red blood cells was suppressed. These results indicate that tea catechins may inhibit the oxidized reduction of vitamin E of the rat body due to unsaturated fatty acids contained in the diet.

5. 2. PREVENTION OF OXIDATION IN PROCESSED FOODS

Many processed fish products, including shellfish, have a high content of unsaturated fatty acids. These unsaturated fatty acids are oxidized by oxygen in the atmosphere, undergo changes in color and taste and develop a strong fishy smell.

5. 2. 1. Anti-oxidation of foods

To prevent the oxidation of these processed foods, some strong antioxidants such as vitamins C or E are used as food additives. Tea catechins also have a strong antioxidant activity as described above, and furthermore are non-toxic. Since they are water-soluble and can permeate foods, they have come to be used as an antoioxidant in processed fish products.

For use in such products green tea catechins are dissolved in water to make a solution. This solution is then sprayed directly onto processed foods in general; and for processed fish products it is made into an oil-soluble product and mixed directly with oils and fats. Tea catechins have an oxidation-preventive activity against the deterioration in the quality of processed foods which is due to gradual oxidation. The development of offensive fishy odors in dried horse-mackerel or salted samon fillets resulting form deterioration by oxidation is prevented by dipping them in a salt solution containing 0.05% tea catechins for several hours.

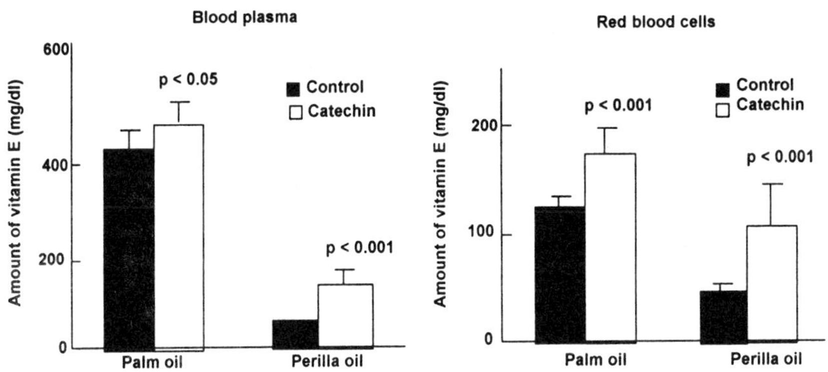

Figure 5. 2. Preventive effect of catechins on oxidation of fish oil-containing DHA. (Kawakami, M., *et al.*, 1996).

Recently, docosahexaenoic acid (DHA) and eicopentaenoic acid (EPA) have been reported to have various beneficial physiological functions, and these "healthy" fish oils have come to be used in a variety of foods. On examination of the suppressive effect of tea catechins on the oxidation of DHA, it was found that the addition of tea catechins extended the time before oxidation occurred, as shown in Figure 5. 2.

5. 2. 2. Preservation of color in foods

Many chemically synthesized food additives are used to make food products look fresh and attractive. Among these some are thought to have toxic effects on human health and physiological functions. On the other hand, the majority of natural plant-derived pigments are not toxic and thus are regarded as safe with regard to human health. So while it is desirable to use natural pigments, they have poor stability as compared with chemically synthesized substances, and Vitamins C or E are usually added to prevent color fading.

With these factors in mind, tea catechins were examined for their effectiveness in preventing fading of color pigments. In the experiments, a β-carotene drink containing carrot juice was left to stand for a month with or without the addition of tea catechins, in a place exposed to sunlight, and the fading of color was compared. As shown in Figure 5. 3, there was a complete loss of color in the β-carotene drink not containing tea catechins, whereas the addition of tea catechins was effective in preserving the color of β-carotene. When the tea catechins were examined to determine which were most effective, it was found that EGCg, EGC and ECg had strong suppressive effects against the color loss of β-carotene.

Similar experiments have been carried out on the color fading preventive effects of tea catechins on other pigments. It was found that tea catechins prevented the fading of natural colors resulting from exposure to light in gardenias, safflowers, cochineal, and monascus. They also prevented deterioration of chlorophyll and riboflavin.

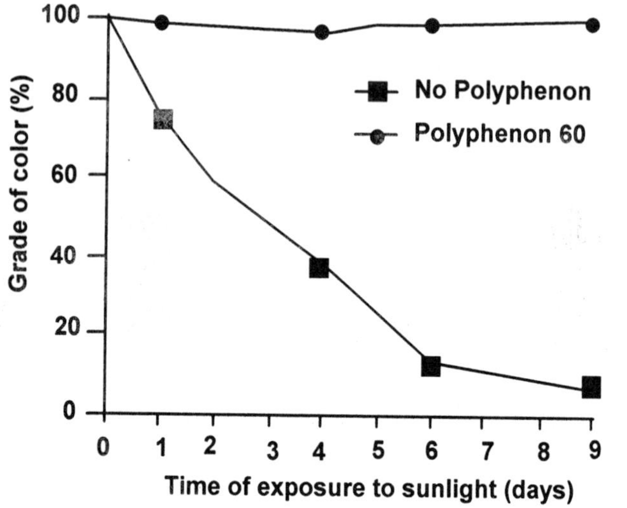

Figure 5. 3. Effect of tea polyphenonls on suppression of color loss of β-calotene. (Shiroishi, S. and Hara, Y., 1992).

5. 2. 3. Deodorant action of unsavory meat flavors

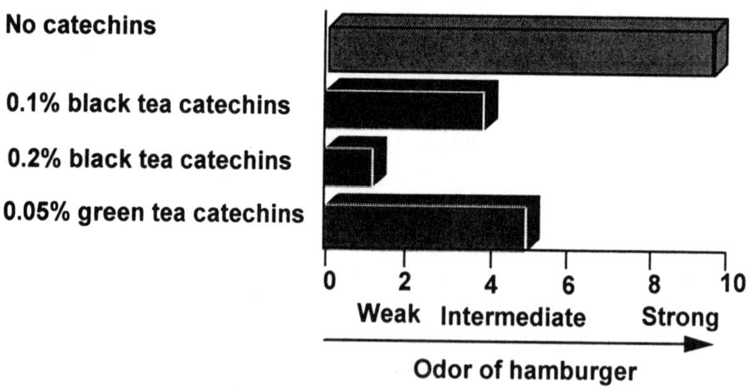

Figure 5. 4. Deodorazing effect of tea polyphenons on the unsavory odors of uncooked hamburger meat. (Kawakami, et al., 1996).

Tea catechins have a deodorant activity on unsavory odors which are characteristic of livestock meat such as beef and pork. As a typical example, tea catechins may be used as a deodorant agent in hamburgers commercially available as frozen foods. When hamburgers are cooked, characteristic and unsavory meat odors develop that may spoil the good flavor. In some local districts, tea is traditionally used in such meals that use meat. With this in mind, polyphonols obtained from green tea and black tea were added to hamburgers and the deodorant effect was examined. As shown in Figure 5. 4, polyphenols from both teas had activity which reduced unsavory meat odors.

5. 3. PREVENTION OF OFFENSIVE ODORS BY TEA

5. 3. 1. Prevention of odors in the home

Tea has a preventive effect against odors that occur in daily life, such as bad breath body odor, and odors that occur in the home, in places such as the refrigerator, bath, or drain in the kitchen. Body odor is due to the degradation of proteins and fatty acids by proliferation of microorganisms in the sweat of the armpit, producing an odorous compound. The stench of sweaty socks is also produced by a similar process, and the odors present in the refrigerator, bath and drain in the kitchen are also due to the proliferation of microorganisms.

Tea catechins eliminate the cause of the odor by binding to the substances produced by microorganisms, and furthermore they have an antibacterial and antibacteriocidal activity against microorganisms. Practically, how can we use tea catechins to prevent these odors? To prevent body odor, absorbent cotton soaked in strong green tea may be used to wipe under the arms after taking a bath or shower. To prevent the odor of socks, after washing they could be soaked in tea and then wrung out and dried.

The smell of the diapers of babies and elderly bed-ridden persons can be a problem for family members. Especially in the case of elderly persons, the strong odor can be very unpleasant. Diapers could be dipped in tea and dried before use, and when changing diapers a cloth dipped in warm tea could be used for wiping. This would result in suppression of the propagation of microorganisms and prevention of odors.

To prevent odors in the refrigerator and bathroom, a cloth bag containing used tea leaves could be placed in a corner of the refrigerator or dipped in the bath.

5. 4. AIR CLEANING ACTION OF TEA

People who suffer from headaches, dizziness, nausea, watery eyes, or rashes from being in a new condominium or office are not uncommon. These symptoms are caused by new building materials, especially adhesives used for plywood and the formaldehyde in wallpaper. When a newly built room is left all closed up for a long time, it becomes full of the formaldehyde released from these building materials, so that anyone who enters the room is likely to suffer from dizziness and nausea..

Tea catechins react well with formaldehyde in the air, and absorb it. Formaldehyde and tea catechins were mixed together in water in a glass tube and left to stand at room temperature for 15 min. Then the amount of formaldehyde in the glass tube was determined. As shown in Table 5. 1, among various plant catechins, green tea catechins most effectively absorbed formaldehyde, and black tea catechins also had a strong absorbing activity for formaldehyde.

Tea catechins were soaked into a filter and left to stand exposed to formaldehyde in a closed glass vessel for 24 hours. The amount of formaldehyde absorbed in the tea catechin-treated filters was determined. Results are shown in Figure 5. 5. A filter treated with tea catechins absorbed several times more formaldehyde than a filter without tea catechins.

Both filters were removed from the glass vessel and left to stand in the room for 24 hours to determine the amount of formaldehyde remaining in these filters. As shown in Fig. 5.5, the amount of formaldehyde released into the air again from the tea catechin treated filters was much less than from the filters not treated with catechins.

Table 5.1. Absorption rate (%) of various plant extracts (catechins) for formaldehyde (Takagaki, A., *et al.*, 2000)

Plant extrcts	Decreasing rate of formaldehyde
Green tea extract	98. 71
Black tea extract	87. 71
Keplacco extract	24. 29
Chestnut extract	19. 44
Mimosa extract	9. 98

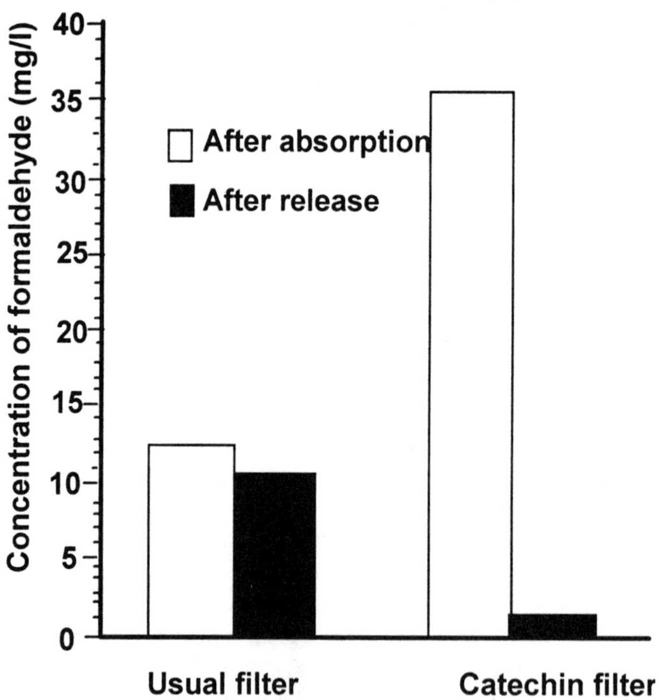

Figure 5. 5. Efdfect of tea catechins on absorption of formaldehyde. (Takagaki, A., *et al.*, 2000).

Recently, air cleaners equipped with tea catechin-treated filters have become commercially available. These filters are effective in absorbing the influenza virus, cigarette smoke, Japanese cedar pollen and formaldehyde released from new building materials. Air cleaners fitted with these filters, specifically for use in the toilet, are commercially available. If these air cleaners are set on the side of the toilet bowl, odors both inside the toilet bowl and in the toilet room may be effectively absorbed.

6

HOW TO MAKE A GOOD CUP OF GREEN TEA

It is a combination of many factors and conditions that determines how good the cup of green tea you drink is. First of all the water used, the temperature it is at when poured onto the tea leaves, the length of time the tea is left to brew; even the way the tea is poured; all these factors may affect the taste and aroma of tea.

6. 1. SUITABLE WATER FOR GREEN TEA

Japan, has an abundance of good tasting water in nature. Japanese who have been abroad may really appreciate how good water in Japan is. Pleasant tasting water, while good for drinking may not necessarily be appreciated for other purposes. There may be differences in the water that is most suitable for making tea and coffee, cooking, making "sake" or for mixing in drinks such as whisky and water.

In general, water that is good for drinking, making tea, and cooking, is soft water deficient in Mg and Ca ions. Water suitable for making "sake" and mixing with whisky is hard water containing only a small amount of these ions. Water suitable for tea, especially for green tea, has no odor and is good for drinking as it is.

Spring water wells up from sources in steep mountains and deep valleys. In Japan, as in other parts of the world, the names of places which can boast these clean clear springs have become renowned due to the bottling of this water for commercial purpose.

However, it is most likely to be tap water that is used for making tea at home on a daily basis. In this case the water should be boiled at 100°C to remove chlorinated limes and left to stand for a few minutes. By doing this calcium, magnesium and sodium salts are removed, odor is eliminated, and the water is made suitable for brewing green tea.

In the tea ceremony an iron kettle is often used to boil water for brewing powdered tea. However, for extracting the regular green tea drunk in the home, an enamel anodized aluminum or aluminum kettle is good for boiling the water. The best temperature for brewing the tea differs according to the kind of tea. For Gyokuro, tea water is boiled once and cooled to about 60°C. For brewing Sencha, the best temperature for the water is 90°C. For coarse tea, roasted green tea, or unpolished rice tea, boiling water is suitable.

6. 2. CHOOSING GREEN TEA

Which tea should be drunk in each situation? This choice is somewhat like choosing wine—it is influenced by the type of food that is to accompany or follow. For example, the green tea that is served in "Sushi" restaurants is a powdered tea with a strong green color but a comparatively light taste, which does not overpower the flavors of various types of fish. Refreshing oolong tea complements Chinese food, which can be quite oily.

Teas differ in taste and aroma according to where they were produced and the procedures used to process them. They cannot simply be evaluated by price alone. Rather it is advisable to choose a tea that suits you both in terms of taste and price.

6. 3. HOW TO POUR A CUP OF GREEN TEA

First, put green tea leaves into a teapot and pour on hot water at the temperature most suitable for the kind of tea used. After leaving the teapot to stand for a while, components of the tea leaves are extract into the hot water, giving the tea a palatable taste and a good aroma. Next, slowly pour a small amount of the tea from the teapot into each of several cups, and when you get to the last cup repeat this process in reverse order, pouring a little at a time until you return to the first cup poured. To obtain the best flavor, make sure the teapot is emptied of every last drop. This procedure is followed to regulate the concentration of tea so that each cup of tea is the same, Each tea cup should be two thirds full or full.

6. 4. KINDS OF GREEN TEA

Gyokuro: After pouring 60°C hot water into the teapot, it is better to allow it to stand stationary without shaking. At this temperature of the hot water, only the amino acids in the tealeaves are extracted, and this is what gives the tea a good flavor. It is also just the right temperature for drinking. For the second extraction, the water should be at a higher temperature, and it should be left to stand for 70-80 seconds.

Sencha: In the case of Sencha, the temperature of the water should be about 90°C. After allowing it to stand for one minute, the tea extract is poured into the cups. At this temperature, not only the amino acids but also the catechins contained in the tea leaves are extracted into the hot water. Catechins have various beneficial effects on human health such as preventing diseases, including life style-related diseases, as described in previous chapters. If the tea extraction is left to stand for too long, catechins continue to dissolve into the extraction, making it taste astringent. An extraction that has a good balance between a moderate amount of amino acids and catechins is best for drinking.

Bancha: Bancha is made from hard tea leaves and stalks so hot water should be used. The taste is light and plain so it is good to drink after meals. This tea is especially recommended for older persons and children.

Houjicha (roasted tea): Houjicha is produced by parching Bancha or Kukicha over a strong heat so that it attains a fragrant aroma. Tea that has been exposed to moisture may have its original good flavor restored by roasting it. As with Bancha, Houjicha requires plenty of hot water and is brewed in the same way, allow it to stand for about 30 seconds before pouring it into the cups. Hot water must be used or the taste of Houjicha will not be good. The tea extract from Houjicha may be chilled for a lovely summer drink.

Kukicha (stalk tea): Kukicha is also known as Bocha, and is produced from the stems, stalks and hard leaves that are removed when Gyokuro and Sencha are produced. Kukicha has a flavor similar to that of high quality Sencha, with a slightly grassy aroma. Brewing can be either with hot water as for Gyokuro or upper quality Sencha, or with water at a lower temperature as is used for Bancha, thus producing two quite different tastes.

Konacha (powdered tea): This tea is known as "Oagari" (green tea) in "Sushi" restaurants. Konacha is brewed by pouring hot water onto the powdered tea, and then filtering the mixture through a tea strainer. A strong green color appears and this tea goes well with "sushi" dishes.

6. 5. TABOOS IN TEA DRINKING

Tea has a tremendous beneficial effect on human health and diseases. There are, however, some taboos about drinking tea.

Overnight tea: It is an old saying that tea should not be made with used tea leaves from the day before. When tea leaves are extracted with hot water, large amounts of proteins and tannin still remain in the leaves. When these tea leaves are left to stand overnight, they may rot, mildew, or be oxidized producing strong stimulatory substances. If hot water is poured onto such tea leaves, these substances are extracted into the hot water and strongly stimulate the stomach and intestine, which can lead to inflammation of these organs. Tea made from tea leaves left overnight is not only second-rate in flavor but may also be harmful to the health.

Tea drinking when hungry: It is not good to drink a large amount of tea on an empty stomach. Drinking tea when hungry may temporarily divert the hungry feeling, but it does not last. When a lot of strong tea is drunk on an empty stomach, a large amount of tannin enters the digestive organs, inhibiting the secretion of gastric juice, and this may produce gastric catarrh.

Tea drinking immediately before going to bed: Tea contains a certain amount of caffeine. If we drink many cups of upper grade tea after dinner, caffeine remains in our system for several hours. Caffeine stimulates the nervous system and we have difficulty falling asleep.

7

FOOD AND INDUSTRIAL APPLICATIONS OF TEA CATECHINS

Looking back over the years it is clear that the ways in which tea is involved in our daily lives have undergone many changes. In the past, tea was purchased exclusively at a shop specializing in teas and drunk at home after brewing it in a small teapot or a kettle. Recently, tea has become available almost everywhere and is supplied commercially in various forms including tea bags, bottles and cans. The products are also varied with tea being mixed with other fruit and vegetable juices.

The most marked changes in the tea industry over the last decade have stemmed from the production of tea drinks packed in cans and bottles. There is no need for a tea cup, tea leaves or teapot, and these ready-to-drink products in their various forms may be purchased from convenience stores and vending machines for consumption anywhere, anytime. One reason for these drastic changes in the industry may be the increasingly convenience-oriented society in which we live, coupled with heightened perceptions about the health benefits tea catechins and polyphenols may impact.

The recent dramatic growth of tea drinks in the soft drink industry is shown in Figure 7. 1.

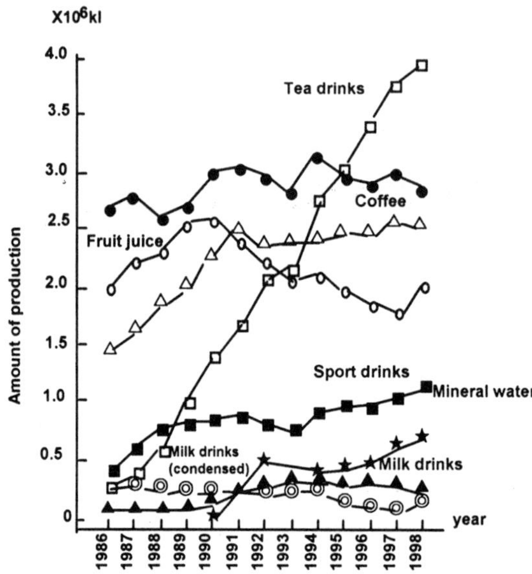

Figure 7. 1. Changes in the amount of soft drinks produced in Japan over the last ten years.

In addition to catechins, caffeine, theanine, saponin, inositol glycoside, vitamin C, vitamin E, potassium, fluorine and many other substances are contained in tea. Over the past two decades, we have concentrated our focus on catechins, which are the main components of tea, and are quite astringent. Crude fractions of tea catechins have been extracted and each individual fraction identified and examined for its medicinal and physiological functions.

Results have shown that tea catechins have excellent anti-oxidant activity, which could possibly help to prevent various life style-related diseases such as cancer, high-blood pressure, high-blood sugar and hyperlipidemia. Tea catechins are also effective in preventing diseases caused by bacteria and virus infections.

While catechins have tremendously beneficial effects on human health and diseases, there are some problems when it comes to making use of them in commercial products. Their astringency, and color instability which is apparent in the liquid state when a color change from green to brown occurs, can complicate the process of manufacturing products containing catechins. Consequently the production cost becomes expensive, even if the original catechin materials used are reasonably priced.

Studies now in progress aim to find suitable vehicles by which the functions of tea catechins can be put to their best advantage, and thus gain worldwide promotion. With this in mind various products of tea catechins that have practical and industrial applications have been developed and are described below.

7. 1. CATECHIN CAPSULES (FRONTISPIECE)

This capsule encompasses all the beneficial effects of tea catechins for the maintenance of good health and prevention of many diseases. In particular it is markedly effective in improving the gastric-intestinal environment and reducing the offensive odor of feces. Some of these catechin capsules are combined with some vitamins for the purpose of inactivating active oxygen radicals.

Others are combined with oligosaccharides, which help to maintain the intestinal flora in good condition.

7. 2. CATECHIN EGGS (FRONTISPIECE)

In experiments with rats and clinical studies with human subjects it was confirmed that tea catechins suppress the levels of cholesterol in the blood and the oxidation thereof, thus playing an important role in the prevention of heart disease. Application of this beneficial function has been carried out by mixing tea catechins in the feed of laying chickens. The resulting eggs contain a reduced amount of lipids and are commercially marketed as "catechin eggs". Further applications of tea catechins, such as the marketing of chickens that have been fed catechins, are new developments in the livestock breeding industry.

7. 3. CATECHIN CANDY (FRONTISPIECE)

Tea catechins can also be effective in the area of oral health by preventing tooth decay and infections with influenza virus. "Catechin candy", which contains tea catechins, has entered the market.

7. 4. CATECHIN TABLETS FOR BAD BREATH

Tea catechins absorb and lessen the odor of fish and garlic. They also have a preventive action on abnormal fermentation in the digestive track. Drinking tea after meals not only refreshes the mouth, but it is expected that catechins will also exert beneficial effects. Catechin tablets for prevention of bad breath are being marketed.

7. 5. CATECHIN DRINKS WITH RADICAL SCAVENGING POWER

Tea catechins, in combination with vitamins such as carotenoid (precursor of vitamin A), have a strong scavenging action on active oxygen. Due to their health benefits these catechin drinks have some avid followers.

7. 6. CATECHIN FILTER (FRONTISPIECE)

When an air purifier fitted with a filter treated with tea catchins is used, all viruses in the atmosphere in the room are completely absorbed into the filter and inactivated, thus easily establishing a virus-free environment. Other applications of the catechin filter include combination with a Hepa filter, use in a futon drier, and use in a facial mask, which is effective in preventing bacterial and viral infections.

7. 7. CATECHIN COSMETICS

Tea catechins have protective effects against UV damage and prevent pigment coloration. Cosmetics and soaps containing tea catechins are now being produced and many more applications in this field are anticipated.

7. 8. CATECHIN MEDICAL PRODUCTS

It is hoped that tea catechins may come to be used as actual medicinal products, not just as health products for prevention of various diseases. However, even extracted tea catechins with a high purity contain a number of minor substances. It is not easy therefore to maintain a constant amount of each catechin in the product. To obtain permission to produce medicinal products, it is necessary to carry out safety tests. These must be conducted thoroughly, and so much expense and time is entailed. Clinical tests for these products can also be complicated. By dealing with these difficulties one by one, it is hoped that medicinal products utilizing catechins will eventually come into worldwide use.

7. 9. PRODUCTS FOR MAINTAINING FRESHNESS OF FISHERY PRODUCTS (FRONTISPIECE)

Catechins inhibit oxidization of the reddish pigment of fish (carotenoid), and also prevent deterioration in the quality of the oil, proliferation of bacteria, and development of fish odor. These activities are of great commercial value, and catechin solutions have been produced for this purpose. The addition of small amounts of this catechin product during the processing stage of preserving dried fish, or to the salt solution, may suppress acidification and the fishy odor of dried fish.

7. 10. DEODORANT SPRAY FOR KITCHENS

Tea catechins have the ability to absorb amines, which are responsible for characteristic fishy smells. They also have an anti-bacterial activity against food poisoning bacteria such as *Vibrio,* which can be found on fish. An ethyl alcohol solution is commonly used in anti-bacterial kitchen sprays. The addition of tea catechins to such sprays makes them effective both for antibacterial and deodorant purposes. These sprays are now being marketed commercially.

7. 11. CATECHIN-CONTAINING ZINC TABLETS

Zinc is a trace metal, which is apt to be lacking in the human body. It is an essential element since it raises the enzyme activity that suppresses superoxide. These tablets combine the strong suppressive action of tea catechins on super oxide with the ability of zinc to vitalize cells in the body.

7. 12. SOFT DRINKS CONTAINING CATECHINS

These healthy soft drinks are made up of a concentrated extract of tea catechins mixed with other plant extracts. As more healthy effects of tea catechins come to light, it is expected that these products will gain a strong standing in the market.

7. 13. CATECHIN-CONTAINING INSTANT GREEN TEA

Instant green tea is made by crushing the water soluble components of green tea into a powder. Catechin powder is added to give the tea a catechin content double that contained in ordinary green tea. The increased astringency due to the large amount of catechins is countered by the addition of Matcha, powdered tea. This product is enjoying popularity among many people who drink it on a daily basis.

7. 14. CATECHIN TEA BAGS

Positive health effects of catechins, and their preventive properties against various diseases, have become known in many countries including the United States. Polyphenons (green tea catechins) mixed with a variety of other health-promoting natural ingredients are being marketed in capsule form on a large scale. While black tea bags are familiar in daily life in western societics, green tea bags are still a novelty and have yet to reach a high degree of acceptance in the market. In order to make them more acceptable to the western market, green tea bags flavored with lemon, mint or berry flavors have been developed.

7. 15. SLUG REPELLANT SPRAY

Catechins have protein-binding activity. The astringent taste of tea may be due to this activity. The body surface of mollusks such as slugs is covered with mucoidal protein. When tea catechins are sprayed on a slug, its body surface is covered with catechins; the slug then has difficulty in carrying out skin respiration and dies. This specific activity of catechins is utilzed in a slug repellant spray, now being sold in gardening shops, that kills slugs without contaminating the soil.

7. 16. CLOTHES WITH ANTIBACTERIAL AND ANTIOXIDANT PROPERTIES

Recently progress has been made in finding ways of utilizing the binding properties of tea in the clothing field. Tea catechins can be absorbed into high molecular weight fibers, producing cloth with strong antibacterial properties. Cloths with antioxidant properties, or which protect against UV rays, are also beginning to come onto the market.

7. 17. FORMALDEHYDE ABSORPTION BY CATECHINS

Tea catechins react with formaldehyde that is released from structural and adhesive materials used in the construction of new housing, causing the so-called "sick house syndrome". Paper and cloth permeated with tea catechins and left hanging in a house in which the level of formaldehyde is high may absorb formaldehyde. Once absorbed, the formaldehyde is not released again even at higher temperatures. The most effective way to utilize tea catechins for the absorption of formaldehyde in new houses is now under study.

Examples of food and industrial applications of tea catechins have been described above.

Research into the health effects of tea catechins continues to progress and so more application products of tea catechins may be seen in the future.

8

CONCLUDING REMARKS

It has long been considered that tea is good for health. However, it is clear from research over the last two decades, that the extent to which tea has had a beneficial influence on human health has been remarkable. Green tea has a long history in Japan, and the act of its consumption by Japanese people has had cultural and spiritual significance, with an affinity to the Buddhist philosophy of "wabi" (spiritual solitude) and "sabi" (element simplicity). At the same time it came to be the most commonly consumed beverage in Japan.

Traditionally, Japanese food culture is dominated by fish and vegetable dishes such as the well-known "Sushi" (fish and rice rolled in sea weed) and "Sashimi" (sliced raw fish), and vegetarian dishes that are arranged skillfully to please the eye as well as the palette. In countries such as the United States and in Europe, where meat is commonly eaten, these kinds of foods are being regarded with growing interest. Japan has the longest life expectancy in the world, and this is thought to be largely due to eating habits, while the role played by tea cannot be overlooked.

One of the good things about tea is that apart from some high grades, it may be purchased at a reasonable price, and requires only the addition of hot water. This book documents the effects of tea based on the results and evidence of scientific studies. Nowadays various kinds of drinks such as fruit juice, coffee, cocoa, carbonated drinks, and lactic acid drinks are sold in convenience stores and automatic vending machines. From this we may be led to think that the children and young people of today do not drink as much tea as was consumed by their counterparts in the past. It appears, however, that this is not true. As shown in Table 8, the total amount of tea consumed, including green tea, has not changed or has gradually increased over the years.

With the changing times, all aspects of Japanese daily life, food, clothing and housing have undergone transformation.

Traditional Japanese clothes have been replaced with European clothes, there has been urbanization of the rural community, mechanization of work formerly done by hand, and the inclusion of more and more Western food into the diet. Under such circumstances, it is necessary to take another look at the effectiveness of tea and how to incorporate it into these modern times. In Japan, the number of elderly people making up the population is on the increase, but it is hoped that a heightened health consciousness will lead to a society in which healthy senior citizens are playing an active role.

Table 8. Annual amount of various beverages consumed as compared with green tea (Ministry of General Affairs, Ann. Rep. Household Budget Survey : Every household throughout Japan)

Year	Total tea	Green tea	Black tea	Other teas	Coffee and cocoa	other drinks
1965	1,804	1,585	219		769	3,883
1975	5,718	5,196	522		3,270	17.265
1985	7,892	6,550	1,342		7,105	19,422
1989	7,544	6,447	1,098		7,140	22,917
1990	7,169	6,215	954		7,235	24,708
1991	7,553	6,585	968		7,340	24,994
1992	7,782	6,807	975		7,355	25,202
1993	8,023	7,131	892		7,549	24,460
1994	7.450	6,661	789		7,572	27,800
1995	11,419	6,699	788	3,932	7,712	23,349
1996	11,912	6,901	813	4,198	8,079	22,827
1997	12,428	7,083	769	4,577	7,823	23,731

Year	Juice	Carbonated drinks	Lactic drinks	Other drinks	Real income	Consumers price index
1965	872	——	1,540	1,471	65,141	25.0
1975	4,197	1,502	5,745	5,821	236,152	55.3
1985	7,908	3,171	4,064	4,290	444,846	87.4
1989	10,546	2,748	4,204	5,419	495,849	90.7
1990	11,739	2,696	4,197	6,076	512,757	93.5
1991	11,772	2,722	4,174	6,325	548,769	96.5
1992	11,616	2,734	4,127	6,725	563,855	98.1
1993	11,091	2,527	4,227	6,835	570,545	99.4
1994	12.516	2,963	3,427	6,394	567,174	100.1
1995	11.574	2,849	3,583	5,343	570,818	100.2
1996	10,981	2,941	3,552	5,353	579,461	100.1
1997	11,016	3,018	3,528	6,170	595,214	101.9

REFERENCES

Doll, R.., 1978, An epidemiological perspective of the biology of cancer., *Cancer Res.*, **38**, 3573-3583.

Fearon, E. R. and Vogelstein, B, 1990, A genetic model for colorectal tumorignesis., *Cell*, **61**, 759-767.

Fukuyo, M., Hara, Y. and Muramatsu,. K., 1986, Effect of tea leaf catechin, (-)-epigallocatechin gallate on plasma cholesterol level in rats. *Nippon Eiyo Shokuryo Gakkaisi (J. Japan. Soc. Nutr. Food Sci.)*, **39:** 495-500 (in Japanese with English summary).

Gupta, S., Hastak, K., Ahmad, N., Lewin, J. S. and Mukhtar, H., 2001, Inhibition of prostate carcinogenesis in TRANP mice by oral infusion of green tea polyphenols, *Proc. Nat. Acad. Sci. USA*, **98**, 10350-10355.

Handa, O., Naito, Y., Takagi, T., Ishikawa, T., Matsumoto, N., Kassai, K., Yoshida, N. and Yoshikawa, T., 2002, Effects of functional ingredients of O-CHA on gastric mucosal inflammatory response against *Hericobacter pylori*, *Proc. Internat. O-CHA Res. Conference*, **16**, 140.

Hara., Y, 2000, Health benefits of green tea catechins. In: *Improvement of Intestinal Conditions in Health Benefits, Physiological Effects, and Chemistry*, edited by T. H. Parliament,, C-T.Ho, and P. Shieberle, Amer. Chem Soc., pp. 165-176.

Hara, Y., 2003, Clinical chemoprevention trials with tea catechins in the USA, *Proc. Joint Symp. World Green Tea Assoc,-Univ. Shizuoka COE21, Tea Health Sci.*, 50-53.

Hara, Y. and Honda, M., 1990:, The inhibition of α-amylase by tea polyphenols., *Agric. Biol. Chem.*, **54**, 1939-1945 .

Hara, Y. and Ishigami, T., 1989, Antibacterial activities of tea polyphenols against food born pathogenic bacteria (Studies on antibacterial effects of tea polyphenols. Part III), *Nippon Shokuhin Kogyo Gakkaishi (J. Japan Soc. Food Sci. Technol.)*, **36**, 996-999 (in Japanese with English summary).

Hara, Y. and Tono-oka, F., 1990, Hypotensive effect of tea catechins in blood pressure of rats. *Nippon Eiyo Shokuryo Gakkaishi (J. Japan Soc. Nutr. Food Sci.)*, **43**, 345-348 (in Japanese with English Summary) ..

Hara, Y., Watanabe, M. and Sakaguchi, G., 1989, The fate of *Clostridum botulimum* spores inoculated into tea drinks. *Nippon Shokuin Kogyo Gakkaishi (J. Japan, Soc. Food Sci. Technol.*), **36**,,375-379 (in Japanese with English summary).

Hattori, M., Kusumoto, I. T., Namba, T., Ishigami, T. and Hara, Y. , 1990, Effects of tea polyphenols on glucan synthesis by glucosyl transferases from *Streptococcus mutans*, *Chem. Pharm. Bull.* **38**, 717-720 ..

Hayakawa, S., Kimura, T., Saeki, K., Koyama, Y., Aoyagi, Y., Noro, T., Nakamura, Y. and Tsumura, M., 2001, Apoptosis-inducing activity of high molecular weight fractions of tea extracts, *Biosci. Biotechnol. Biochem.*, **65**, 459-462.

Hibasami, H., Achiwa, Y., Fujikawa, T. and Komiya, T., 1996, Induction of programmed cell death (apoptosis) in human lymphoid cells by catechiun compounds, *Anticancer Res.*, **16**, 1943-1946.

Hirayama, Y..,1978, Prevention of Cancer (in Japanese), p.42, Kodansha, Tokyo.

Honda, M. and Hara, Y., 1993, Inhibition of rat small intestinal sucrose and a-glycosidase activities by tea polyphenols. *Biosci. Biotech. Biochem.*, **57**, 123-124 .

Horiuchi, Y., Toda, M., Okubo, S., Hara, Y. and Shimamura, T., 1992, Protective activity of tea and catechins against *Bordetella pertussis*, *Kansenshogaku Zasshi (J. Japan. Assoc. Infect. Dis.*), **66**, 599-605 (in Japanese with English summary).

Ishigami, T. and Hara, Y., 1993, Proc. Internat.. Symp. Tea Sci Tea Sci., Human Health, Calcutta, India, 125.

Ishigaki, A., Tono-oka, F., Matsumoto, N. and Hara, Y, 1991, Suppression of the accumulation of body and liver fat by tea catechins. *Proc. Internat. Symp. Tea Sci.*, 309-313 .

Iwata, K. Inoyama, T., Miwa, S., Kawaguchi, K. and Koike, G., 1991, Effect of oolong tea on plasma lipids and lipoprotein lipase activity in young women., *Nippon Eiyo Shokuryo Gakkaishi (J. Japan. Soc. Nutr. Food Sci.)*, **44**:, 251-259 (in Japanese with English summary).

Kada, T., Kaneko, K., Matsuzaki, S., Matsuzaki, T. and Hara, Y., 1985, Detection and chemical identification of natural bio-antimutagens. A case of the green tea factor., *Mutation Res.*, **150**, 127-132 .

Kanaya, S., Goto, K., Hara, Y. and Hara, Y., 1991, The physiological effects of tea catechins on human volunteers. *Proc. Internat. Symp. Tea Sci.*, 314-317, Shizuoka, Japan.

Katsuno, Y. and Isemura, M., 2002, Apoptosis-inducing activity of the enzyme digestion product of green tea residue, 2002, *Proc. Internat. O-CHA Res. Conference*, **16**, 132.

Kawakami, M., Nanjo, F. and Hara, Y., 1995, Application of green tea extracts to food products. II. *Shokuhin Kogyo (Food Industry)*, **38**, 78-81 (in Japanese).

Kawakami, M., Nanjo, F. and Hara, Y., 1996, Functionality of the tea (No.15). Application of green tea extract "polyphenon" in food (No. 2)., *Nippon Shokuhin Kogyo Gakkaishi (J. Japan. Soc. Food Sci. Technol.)*, **39**, 71-73 ..

Kee, M., 1982, Cancer causation in booklet form. *Nature*, **303**:, 648 (1983).

Kinoshita, T., Kawagishi, H., Suzuki, M., Nanjo, F., Hara, Y. and Furukawa, S., 1994, Stimulating activity of NGF synthesis by teanine and its derivatives, *Abstract Soc. Biosci. Biotech. Agrochem.*, p.203 (in Japanese).

Kondo, T., Ohta, T., Hara, Y., Watanabe, H., Igura, K., Eto, N. and Kaji, K., 2001, Tea catechins inhibit angiogenesis in *in vitro* model system, *Abstract Soc. Biosci. Biotech. Agrochem.*, p. 22 (in Japanese).

Kumagai, I., Sugiyama, K. and Shinmura, J. (eds.), 1999, Green Tea Culture and Japanese. Gyousei, Tokyo (in Japanese).

Kumakura, I., Sugiyama, K., *et al.*, eds., 1999, Green Tea Culture and Japanese. — For World O-CHA Festival —, Gyousei, Co., Ltd., Tokyo.

Kuroda, Y.., 1990, Antimutagenic activity of vitamins in cultured mammalian cells., in: *Antimutagenesis and Anticaricinogenesis Mechanisms II*, edited by Y. Kuroda,, D. M. Shankel and M. D. Waters, (Plenum Press, New York and London), pp. 233-256.

Kuroda, Y., 1995, Mutagens and Antimutagens. In: *Antimutagens, Anticarcinogens and Their Detection Procedure*, edited by Y. Kuroda, (Kodansha Scientific, Tokyo), pp. 1-12.

Kuroda, Y. and Hara, Y., 1999, Antimutagenic and anticarcinogenic activity of tea polyphenols. *Mutation Res.*, **436, :**69-97 ..

Li, N., Chen, X., Liao, J., Yang, G., Wang, S., Josephson, Y., Han, C., Chen, J., Huang, M. T. and Yang, C. S., 2002, Inhibition of 7,12-dimethylbenz[a]anthracene (DMBA)-indiced oral carcinogenesis in hamster by tea and curcumin, *Carcinogenesis*, **23**, 1307-1313.

Matsumoto, N. Tono-oka, F., Ishigaki, A. and Hara, Y., 1991, Reduction of blood glucose level by tea catechins. *Proc. Internat. Symp. Tea Soc., Shizuoka, Japan*, pp. 318-321..

McCormick, D. L., Burns F. J. and Albert, R. E., 1980, Inhibition of rat mammary carcinogenesis by short dietary exposure to retinyl acetate. *Cancer Res.*, **40**, 1140-1143

Morita, K., Hara, Y. and Kada, 1987, Studies on natural desmutagens: Screening for vegetable and fruit factors active in inactivation of mutagenic pyrolysis products from amino acids, *Agricol. Biol. Chem.*, **42**, 1235-1238.

Muramatsu, K., 1991, Summary and prospect of physiological function researches of tea. In: *Science of Tea*, K. Muramatsu, ed., Asakura-shoten, Tokyo, pp. 205-213. (in Japanese).

Muramatsu, K., Furuyo, M. and Hara, Y.., 1986, Effect of green tea catechins on plasma cholesterol level in cholesterol-fed rats. *J. Nutrit. Sci. Vitaminol.*, **32**, 613-622

Nakamura, Y., 2001, Tumor-inhibiting substances newly found in tea leaves, *Pro. Res. Meet. Establish. Tea Funct. Test Systems Analysis Funct. Tea Catechins* 19-22 (in Japanese).

Nakamura, Y., Hirasawa, A., Imada, M., Kuroyanagi, M. and Noro, Y., 2002, Inhibition of carcinogenesis by tea non-dialysates of tea leaves, *Proc. 17th Tea Sci. Res. Meeting*, 57-60 (in Japanese).

Nakane, H.., Hara, Y. and Ono, K.:, 1994, Tea polyphenols as a novel class of inhibitors for human immunodeficiency virus reverse transcriptase. *Food Phytocehmicals. II. : Teas, Spices and Haerbs*, Amer. Chem. Soc.., pp. 56-64.

Nakayama, M., Suzuki, K., Toda, M., Okubo, S., Hara, Y. and Shimamura, T., 1990, Inhibition of the infectivity of influenza virus by tea polyphenolss, *Antiviral Res.*, **21**,:289-299.

Nanjo, F., Honda, M., Okusho, K. Matsumoto, N., Ishigaki, F., Ishigami, T. and Hara, Y., 1993, Effects of dietary tea catechins on a-tocopherol levels, lipid peroxidation and erythrocyte deformability in rats fed on high palm oil and perilla oil diets. *Biol. Pharm. Bull.*, **16**, 1156-1159 ..

Oguni, I. and Hara, Y.., 1991, Tea is Very Effective for Cancer and Adult Diseases,, Chunichi News Paper Co. Ltd., Nagoya (in Japanese), pp. 239.

Oguni, I., Nasu, K., Kanaya, S., Ota, Y., Yamamoto, S. and Komura, T.., 1989, Epidemiological and experimental studies on the antitumor activity by green tea extracts, *Japan J. Nutrion*, **47**, :93-102 ..

Oishi, S.., ed.,1988 , Tea in Japan. Tea and Its Production (in Japanese), Gyousei, Tokyo , p. 21..

Oku, N., Yamakawa, S., Asai, T., Uchida, T., Matsukawa, M. and Akizawa, T., 2003, Anti-angiogenic effect of green tea polyphenol through inhibition of membrane-type 1 matrix metalloproteinase, MT1-MMP, Proc. *Joint Symp. World Green Tea Assoc.-Univ. Shizuoka COE 21, Tea Health Sci.*, 26-33.

Sadakata, S., Fukao, A., Hisamichi, S., 1992, Martality among female practitioners of Chanoyu (Japanese 'tea-ceremony'), Tohoku J. Exp. Med., 166, 475-477.

Sakai, M. and Hara, Y., 1995, Inhibition of saccharine digestive enzymes by tea catechins. *Nippon Shokuhin Kogyo Gakkaisi (J. Japan. Soc Food Sci.,. (J. Japan Soc., Food Sci. Technol.)*, **38**,77-81 (in Japanese) ..

Shimamura, T. and Hara, Y., 1996, Method of strengthening, Abstract. In: *Action of Antibiotics, Australia Patent No.* 685029 (1996).

Shiroishi, S. and Hara, Y., 1992, Antioxidative effect of brown catechins and their application. *Nippon Shokuhin Kogyo Gakkaishi (J. Japan. Soc. Food Sci.Technol.)*, **35**,:34-39 (in Japanese with English summary) ..

Sugatani, J., Iwai, C., Yoshinari, K., Abe, I., Noguchi, H. and Miwa, M., 2003, Antiallergic effects of tea components, *Proc. Joint Symp. World Green Tea Assoc.-Univ. Shizuoka COE21, Tea Health Sci.*, 71-75.

Takagaki, A., Fukai, K., Nanjo, F. and Hara, Y.., 2000, Reactivity of green tea catechins with formaldehyde, *J. Wood Sci.* **46**,:334-338 (in Japanese).

Takebe, H.., 1991, Why, Is Cancer Produced ? (in Japanese), Shokabo, Tokyo , p. 66..

Toyoshima, K., 1991, Background of oncogenes and anti-oncogenes, in *Oncogenes and Anti-oncogenes*, edited by M. Shibuya,, (TokyoUniv. PressTokyo), p. 6. (in Japanese).

Wynder, E. L. and Gori, E. G. B., 1977, Contribution of the environment to cancer incidence: An epidemiologic exercise. *J. Nat. Cancer Inst.*, **58**, 825 - 832 .

Yang,, C. S., Maliakal, P. and Meng, X., 2002, Inhibition of carcinogenesis by tea, *Ann. Rev. Pharmacol. Texicol.*, **42**, 25-54.

Yeisai:, 1988, Tea and Health Promotion (Kitha-Yojoki) (in Japanese), Gyosei, Tokyo, p. 21.

AUTHOR INDEX

Doll, R., 18

Fearon, E. R., 17

Gori, E. G. B., 14

Handa, O., 37
Hara, Y., 31, 43, 44, 46, 49, 52, 65, 67-72, 74, 77, 84, 99
Hayakawa, S., 36
Hibasami, H., 36
Hirayama, Y., 21
Honda, M., 46
Horiuchi, Y., 62

Isemura, M., 36
Ishigaki, A., 50
Ishigami, T., 65, 67, 70
Iwata, K., 55

Kada, T., 24, 28, 29
Katsuno, Y., 36
Kawakami, M., 66, 83, 84
Kee, M., 15
Kinoshita, T., 57
Koanaya, S., 44
Kondo, T., 37
Kumakura, I., 12
Kuroda, Y., 31, 33, 35

Li, N., 36
Lu Yu, 1

Matsumoto, N., 47, 48
McCormic, D., I., 34
Morita, K., 24
Muramatsu, K., 29, 30, 53, 54

Nakamura, Y., 36
Nakane, H., 76
Nakayama, Y., 61
Nannjo, F., 81

Oguni, I., 24-27, 52
Oishi, S., 1, 2
Oku, N., 37

Sadakata, S., 28
Sakai, M., 49
Shimamura, T., 74
Shiroishi, S., 84
Sugatani, J., 38

Takagi, A., 86
Takebe, H., 22
Tei, S., 77
Toyoshioma, K., 16

Vogelstein, B., 17

Watanabe, M., 68
Winder, E. L.

Yang, C. S., 36

SUBJECT INDEX

α -amylase, 46
absorption of mutagen, 33
Acanthopanax, 24
Acapose, as drug for diabetes, 45
ACE: *see* Angiotensin I cinverting enzyme
Actinomyces
    produces diabetes drug, 45
active oxygen, 81
adenoma, 17
Aeromonas,70
    inhibited by tea catechins, 67
afllatoxin B_1, 19, 20, 30
    as mutagen, 32
Agrimonia, 24
air cleaning action, 85-87
alcohol, as cause of cancer, 14, 15
allergens, 38
allergic conjunctivitis, 73
allergic diseases
    prevented by green tea, 75
allergic reaction
    prevented by oolong tea, 75
allergic reactions, 73
allergic rhinitis, 73
alminium kettle, 89
Alzheimer's dusease, 56
    prevented by tea, 57-58
amine, 20
amylase
    inhibited by tea catechins, 4
amyloid protein, 56, 57

angiogenesis, 36
Angiotensin I, 41, 42
angiotensin receptor, 16
Angiotensinogen, 41, 42
Angiotension I converting enzyme, 41,
Angiotension II, 41, 42
angular stomatitis, 12
annual amount of beverages, 100
anteriosclerosis
    induced by LDL cholesterol, 55
anteriosclerosis, 45, 51
anti-allergy, 12
anti-asthma, 12
antibacteria, 12
antibacterial activity, 81-87
antibacterial clothes, 97
antibacterial effect
    of tea catechins, 62, 71
antibacteriocidal activity, 85
antibiotics, 13
anticarcinogenic activity, 23
    of tea catechins, 32
anti-carcinogenic substances, 28
anticarcinogenicity, 12
anticarcinogens, 19, 23
anti-dental caries, 12
antidiabetes, 12
antigen, 38
anti-inflammatory effects, 37
antimutagenic activity
    of tea, 28-30
antimutagenicity, 12
anti-oncogenes, 16, 17
antioxidant clothes, 97
antioxidant, 82
anti-oxidation, 82
antioxidative activity, 81-87
antioxidativity, 12
antiscorbutus, 12
apple, 23
appoptosis
    see programmed cell death
Ardisia, 24
aromatherapy effects, 12
arteriosclerosis, 51-56
    mechanisms of, 51
arylamine, as mutagen, 32
As, 19
asbestos, 19
astringency, 94
astrocytes, 57
atmosphere, 13
atmosphere, as carcinogens, 19
average life span, 12
azidothymidine, 76
azomehtyl methane, as carcinogen, 32
AZT: *see* azidothymidine

Bacillus
    inhibited by tea catechins, 67
Bacillus subtilis, 19, 28, 30, 67, 70
    as test system, 32

bacteria
for mutagenicity test, 28
as test system, 19
bactericidal effect
of green tea, 65
bacteriocidal activity
of tea catechins, 67
Bacteriodaceae, 70, 71
bad breath, 63
bad cholesterol, 51, 52
decreased by tea catechins, 55
Bancha, 9, 29, 30
how to pour, 90
basic fibroblast growth factor, 37
β-calotene, 84
β-carotene drink, 83
bean paste (miso), as carcinogen, 20
beneficial bacteria (good bacteria), 70
benz〔a〕pyrene, 18, 19, 20, 30
as mutagen, 32
as carcinogen, 32
best temperature, 89
bFGF: see basic fibroblast growth factor
Bifidobacterium, 70, 71
bile acid, 51
birth rate, 13
black tea
as antimutagen, 32
consumed in a year, 100
inhibits dental caries, 65
black tea extract, as antimutagen, 32
black tea, 9, 10, 30
black pepper, 23
bleeding
in the eye, 45
blindness, 45
blood pressure, 43
blood sugar level
suppressed by tea catechins, 49,
blood sugar levels, 47
blood vessel, 18
Bombix mori, 19
body odor, 85
brain function, 12
brain hemorrhage, 51
brain strokes, 43
broccoli, 23
buckwheat
inhibits ACE enzyme, 42
Buddhism, 3
burdock, 23

C: *see* catechin, 46
inhibits α-amylase activity, 46
Ca, 19
cabbage, 23
caffeic acid, 23
caffeine
improves mental activity, 56
obstructs falling sleep, 91
stimulates the central nervous system, 56
cancer, 13, 16, 33
cancer, as death cause, 51
cancer cell, 16
cancer death, 28
cancers
inhibited by tea catechins, 32
canine kidney cells, 62
capsicum, 23
carbonated drinks
consumed in a year, 100
carcinogenesis, 12, 13, 17
carcinogenic agents, 18
carcinogenic process, 15
carcinogenic subtances (carcinogens), 28
carcinogens, 13, 19, 20-23,30, 32
in food materials, 20-23
cardiotonic effect, 12
carotene, 35, 82
carrots, 20
caspase 8, 36
cataract, 12
(+)-catechin
chemical structure of, 31
catechin candy, 94
catechin capsules, 94
catechin cosmetics, 95
catechin drinks, 95
catechin eggs, 94
catechin filter, 95
(+)-catechin gallate, 46
catechin medical products, 95
catechin tablets, 95
catechin tea bags, 96
catechins
as antimutagenic substances in green tea, 28
chemical structure of, 31
contained in tea, 94
extend life span, 44
for spontaneous mutations, 29
inhibits α-amylase activity, 47
catechin-treated filter, 87
cell membrane, 17
cellulase, 36,
central nerve, 12
cereals, as high risk foods, 20
cerebral apoplexy, 13
as death cause, 51
cerebral cortex, 56
cerebral hemorrhage, 28

cerebral vessel injury, 71
cerebromeningitis, 73
Ch'a Ching, 1
Cha-no-yu (Japanese tea ceremony), 28
chemical substances, 18
  contained in tea, 29
Chinese hamster cells
  as test system, 32
Chin-Hsin oolong (Taiwan oolong), 75
chlorinated limes, 89
chlorogenic acid, 23
chlorophyll, 83
cholesterol, 41, 51
  in blood plasma, 53
choosing green tea, 89
chromosome aberrations, 30
chromosomes, 15
chronic inflammation, 37
cinnamaldehyde, 23, 24
clinic chemopreventive trials, 38
clinical first phase experiment, 38
clinical second phase experiment, 38
Clostridium, 67, 70, 71
  inhibited by tea catechins, 67
cochineal, 83
coconut oil, 49
coffee and cocoa
  consumed in a year, 100
colon cancer, 16, 17, 22
  inhibited by catechins, 32
colon epithelium, 17
color fading, 83
color in foods, 83
color instability, 94
coltsfoot flower, 19
condyloma, 77, 78
conjunctival membrane, 73
constipation, 72
consumers price index, 100
cosmetic products, as mutagens, 18
coumarin, 23
cultured cells, 30
cultured mammalian cells, 19
  for mutagenicity test, 28
cumarin, 24
curcumin, 23
cycard seeds, 19
cycasin, 19
cysteine, 23
cytoplasm, 17

DCC, 17
death rate
  of stomach cancer, 24
decaffeinated black tea
  as anticarcinogen, 32
decaffeinated green tea
  as anticarcinogen, 32
deficiency, 17
dental caries, 12, 63-67
dental plaque, 63
deodorant action, 84
deodorant spray, 96
deodorization, 12
dermatitis, 12
detoxication enzymes, 23
DHA: *see* docosahexaenoic acid
diabetes
  cause of, 45
  prevention of, 45-47
  tea reduces, 49
diet, as cause of cancer, 14, 15
diethyl benzo [a] anthracene, as carcinogen, 32
diethylnitrosoamine, as carcinogen, 32
dimethyl benzoanthracene, as carcinogen, 32
dioxin, 19, 20
diploma
  in tea ceremony, 7
diuretic effect, 12
Divine Healer, 1
DMBA, as carcinogen, 34
DNA methylation, 17
DNA, 15
docosahexaenoic acid, 83
dorsal sac model mice, 37
dried horse-mackerel, 82
drinks, 13
  as mutagens, 18
Drosophila, 19
duodenal tumors
  inhibited by catechins, 32
eating habits, 23, 25

EC
  effective in condyloma, 77
  inhibits α-amylase activity, 46
  inhibits angiogenesis, 37
  inhibits bacterial enzyme, 64
  inhibits bacterial growth, 67
  see epicatechin
ECg
  effective in condyloma, 77
  inhibits α-amylase activity, 46
  inhibits angiogenesis, 37
  inhibits bacterial enzyme, 64
  inhibits bacterial growth, 67
  inhibits platelet aggregation, 38
  inhibits reverse-transcriptase, 76
  see (-)-epicatechin gallate
  suppresses color loss, 83

EGC
effective in condyloma, 77
inhibits angiogenesis, 37
inhibits α-amylase activity, 46
inhibits bacterial enzyme, 64
inhibits bacterial growth, 67
see (-)-epigallocatechin
suppresses color loss, 83
suppresses blood pressure, 42
EGCg
as anticarcinogen, 32
as antmutagen, 32
decreases LDL cholesterol, 54
diminishes fish odor, 66, 67
effective in condyloma, 77
has antibacterial effct, 62
inhibits angiogenesis, 37
inhibits apoptosis, 36
inhibits bacterial enzyme, 64
inhibits bacterial growth, 67
inhibits platelet aggregation, 38
inhibits reverse-transcriptase, 76
kills MRSA, 73
prevents influenza virus infection, 61
suppresses color loss, 83
see (-)-epigallocatechin gallate
EGF, *see* epidermal growth factor
eggplant, 23
eicopentaenoic acid, 83
embryonic toxicity
of Polyphenon E. 38
endocarditis, 73
endothelial cells, 37
Enterobacteria, 70
Enterobacteriaceae, 70, 71
environmental factors
for carcinogenesis, 14
EPA: *see* eicopentaenoic acid
(-)-epicatechin::*see* EC
as anticarcinogen, 24
chemical structure of, 31
(-)-epicatechin gallate: *see* ECg
as anticarcinogen, 24
chemical structure of, 31
(-)-epigallocatechin,: *see* EGC
as anticarcinogen, 24
chemical structure of, 31
(-)-epigallocatechin gallate, *see* EGCg
as anticarcinogen, 24
chemical structure of, 31
effective in antimutagenicity, 28
epidemiological data, 38
epidermal growth factor, 16
epigallocatechin-4-*o*-methyl gallate, 75
epinephrine, 57
Escherichia coli, 19, 30, 71
as test system, 32
esophageal tumors
inhibited by catechins, 32
ethyl methanesulfonate, as mutagen, 35
ethyl-nitro-nitrosoguanidine, as carcinogen, 32
Eubacteria, 71
exaust fumes, as mutagens, 18
experimental animals, 19

fecal odor, 72
fermentation
of green tea, 9, 29
fermented tea, 9
fertilizer, 79
FGF: *see* fibroblast growth factor, 16
fish and meat, 19
fish smell
diminished by tea catechins, 66, 67
fish, 20
fishery products, 95
fluorine
contained in tea, 94
food, 13, 18
as carcinogens, 19
as mutagens, 18
food additives
as cause of cancer, 14, 15
as mutagens, 18
food color, 20
food fibers, 29
food goal in maze, 56
food materials, 23
food poisoning bacteria, 67
food poisoning, 67-70, 73
forestomach tumors
inhibited by catechins, 32
formaldehyde, 85, 97
fresh vegetables, 20, 21
fruit fly: see Drosophila
fruits, 20, 21, 23
FT1: *see* theaflavin
Fukamushi-cha
suppresses histamine, 75
Fuki-no-tow, 20
fukinotoxin, 19, 20
full-fermented tea, 29
fungi, 19, 20
fungi
for mutagenicity test, 28
as test system, 19
furylfuramide, 20

G: *see* (+)-catechin gallate
inibits α-amylase activity, 46

G protein, 16
gallanan, 57
gallic acid, 23
    as anticarcinogen, 24
    inhibits α-amylase activity, 46
gardenias, 83
gargling, 63
garlic, 23
gastric epithelial cells, 37
gastric mucosa, 37
GC
    inhibits α-amylase activity, 46
GCg
    inhibits α-amylase activity, 46
gene expression, 16
gene mutations, 30
gene therapy, 17
genes, 15
genetic change, 17
genetic information, 15
genetic make-up, 13
genotoxicity
    of Polyphenon E, 38
germicidal action, 73
ginger, 23
glucagon masses, 65
glucose
    produces neural tranmitter,56
Glu-P-1, as mutagen, 32
glutathione, 23
glutony, 20
good cholesterol, 51, 52
    increased by tea catechins, 55
good cup of tea, 89-91
grain, 20
green tea, 9, 24
    as anticarcinogen, 24
    consumed in a year, 100
    cultivation, 26
    for stomach cancer, 25
    inhibits dental caries, 65
    kills bacteria, 68
green tea catechins
    as anticarcinogen, 32
    decrease LDL cholesterol, 53
    delay LDL cholesterol oxidation, 55
    suppress liver fats, 50
green tea drink
    kills bacteria, 69
green tea extract, as antimutagen, 2
green tea infusion, as anticarcinogen, 32
Gyokuro
    best temperature for, 89
    how to pour, 90
    stimulates NGF synthesis, 57
Gyokuro, 28-30

H^3-labeled EGCg, 38
half fermented tea, 9, 29
hamburgers, 84
hamsters, 19
handring
    after storage of tea, 59
hard water, 89
harmful bacteria (bad bacteria), 70
HBGF: *see* heparin-binding growth factor
HDL: *see* high-density lipid cholesterol
heart diseases, 12, 13, 28
    as death cause, 51
heat-inactivation of viruses, 75
heavy smokers, 19
hematopoietic system, 76
hemolysis, 12
hemophilia, 75
heparin-binding growth factor, 16
hepatomas
    inhibited by catechins, 32
hereditary factors
    for carcinogenesis, 13
Helicobacter pylori, 37
heterocyclic amine, as mutagen, 32
Hideyoshi Toyotomi, 4
high blood pressure, 12, 13
high blood sugar, 12, 45
high risk foods, 20, 21
high-density lipid cholesterol, 51
hipocampus, 57
histamine, 75
HIV: *see* human immunodeficienctyvurus
Hojicha
    suppresses histamine, 75
Hojicha, 29
hormone
    as cause of cancer, 14
horseraddish, 23
hot rice gruel, 22
Houjicha
    how to pour, 90
Houshucha, 30
how to pour tea, 90
HPV: *see* human papilloma virus
human blood cholesterol
    tea effect on, 55
human blood pressure, 44
human blood sugar levels, 48
human chromosome 21, 56
human colon cancer cells, 36
human genital organs, 77
human histocytic lymphoma cells, 36
human immunodeficiency virus, 75-76

human influenza, 62
human lymphoid leukemia, 36
human lymphoma cells, 36
human stomach cells, 36
human umbrical vein endothelial cells, 37
HUVEC: *see* human umbrical vein endothelial cells
hyperplasia, 17

IgE, 73
immune response, 38
immune system, 12
inactivation
    of mutagen, 3
incinerators, 18
industrial application of tea, 93
industrial products, as cause of cancer, 15
industrial products, 19
infant death rate, 13
infections
    as cause of cancer, 15
infectious bacteria, 14
infectious diseases, 61-78,
inflammation, 76-78
influenza infection, 12 61-63
influnza virus, 61
inhibition zone, 74
initiation, 16, 17, 19, 28, 30
    of carcinogenesis, 33
initiator
    of carcinogenesis, 33
inositol
    contained in tea, 94
insects, 19
    for mutagenicity test, 28
insomnia, 41
instant green tea, 96,
insulin, 45
intercellular matrix, 51
interleukin, 8, 37
intestinal bacterial flora, 12
intestinal conditions, 72
intra-hospital infections, 72, 73
intra-hospital infective bactria, 72
intraintestinal bacteria, 70-72
ionizing radiation, as cause of cancer, 15
iron kettle, 8
irritation, 41

Japanese pepper, 23
JB6, *see* mouse epidermal cells
juice
    consumed in a year, 100

Kadsura, 24
Kamairi-cha, 30,
    suppresses histamine, 75
Kawane-cha, 26
kempherol, 23
kinds of green tea, 90
Kissa-Yojoki, 3, 11, 56
Konacha
    how to pour, 91
K-ras, 17
Kukicha
    how to pour, 91

lactic acid bacilli, 70
lactic drinks
    consumed in a year, 100
Lactobacillus, 70
Lactobacteria, 71
large intestinal tumors
    inhibited by catechins, 32
large intestine, 17
LDL: *see* low-density lipid cholesterol
life span
    in SHR, 44
life style, as carcinogens, 19
Li-Fraumeni syndrome, 16
lipid peroxidation, 12
lipoproteins, 51
liver cancer, 22
low risk foods, 21
low temperature
    for storage of tea, 59
low-density lipid cholesterol, 52
Lu Yu,, 3
lucidophyllus forest, 1, 2
lung cancer, 18, 19
lung tumors
    inhibited by catechins, 32

malignant cancerous cells, 15
malignant cells, 17
malignant neoplasia, 13
malignat transformation, 17
maltose, 46
mammary gland tumors, 35
    inhibited by catechins, 32
mast cells, 73, 75
master system
    for tea ceremony, 7
Matcha, 3, 29, 30
maximum blood pressure, 44
MDCK: *see* canine kidney cells, 62
mechanisms
    of inhibition of carcinogenesis, 33
medical drugs:, as cause of cancer, 15
medicinal effects, 11
medicinal herb, 24

medicinal plants, 23
medicines, 13
as cause of cancer, 14
as mutagens, 18
meiosis, 16
memory, 57
mental activity
increased by tea, 56
metabolic activation, 33
metastasis
of cancer cells, 34
methicillin
against MRSA, 73, 74
methylbenzanthracene
as carcinogen, 32
mice, 19
milk drinks, 93
milk, 20, 21
mineral water 93
minimun blood pressure, 44
mitomycin C
as carcinogen, 32
Molt 4B cells: *see* human lymphoid leukemia
monascus, 83
monkey chair
as anticarcinogen, 24
motor vehicles, 20
Motoyama-cha, 26
mouse epidermal cells, 36
MRSA, 72, 73
synergistic effects against, 73, 74
multi-step carcinogenesis, 16, 17, 30
mutagen precurcor: *see* promutagen
mutagen, 18, 30, 33
mutagenicity
of Polyphenon E, 38
mutaions of chromosomes, 28
mutant cell, 16, 33
mutations of genes, 28
myocardial infarction, 45

natural spices, 23, 24
nerve cells, 57
nerve function, 12
nerve growth factor, 57
increases NGF secretion, 58
NGF: *see* nerve growth factor
night blindness, 35
nitrisoguanidine, as mutagen, 32
nitrite, 19, 20
nitrosamine, 19, 20
nitrosodiethylamine, as carcinogen, 32
nitrosodimethlamine, as carcinogen, 32
nitrosoguanidine, 30
as mutagen, 32
nitroso methyl benzyl amine, as carcinogen, 32
Nobunaga Oda, 4
N-OH-Trp-P-2, as mutagen, 32
non-fermented green tea, 29
non-mutagen, 33
non-smokers, 18
normal cells, 15, 16, 33
4NQO
as mutagen, 32
see 4-nitroquinoline-1-oxide
nutrient, 12
obesity
tea prevents, 49-51
offensive odors, 85
oligosugars
control digestive tract, 72
Omote-Senke
oncogenes, 16, 17
onion, 20, 23
oolong tea, 9, 30
inhibits dental caries, 65
kills bacteria, 68
prevents allergic reaction, 75
Osuga town (O town), 26. 27
Otemae, 28
overnight tea, 91
oxacillin
against MRSA, 73, 74
oxidation-preventive activity, 82

p53, 17, 18
PAF: *see* platelet-activating factor
palm oil, 82
p-anisaldehyde, 24
papilloma virus, 77
PDGF: *see* platelet-derived growth factor
pectinase, 36
penicillin, 13
peony, as anticarcinogen, 24
perilla oil, 82
pharmacokinetic study
of tea catechins, 38
phenylalanine pyrolysate, 20
Phe-P-1: *see* phenylalanine pyrolysate
phloxine, edible red dye, 20
phospholipid, 52
physiological effects, 11
pickled radish, 79
pickles, 20
pineapple, 23
plant fibers, 23
platelet agglutination, 12
platelet-activating factor, 38
platelet-derived growth factor: *see* PDGF
platelet-derived growth factor, 16

Plesiomonas, 70
  inhibited by tea catechins, 67
pollen allergy, 73, 75
  prevented by green tea, 75
pollution, as cause of cancer, 15
polyphenol, as antimutagen, 32
polyphenol oxidase, 9
polyphenols
  as anticarcinogen, 32
polyphenols, 29
polyphenon
  lowers blood sugar levels, 47, 48
Polyphenon 100, 73
polyphenon 60
  suppresses color loss, 84
  changes intraintestinal bacteria, 71
Polyphenon E
  effective in condyloma, 77
  used for clinical chemopreventive trials, 38
polysaccharides, 45
potetoes, as high risk foods, 20
potassium
  contained in tea, 94
precancerous state, 38
preservative, 20
prevention
  against allergy, 38
preventive effects
  against human diseases, 11
procarcinogen, 23, 33
processed foods, 82-84
profession, as cause of cancer, 15
professional factor, as cause of cancer
programmed cell death, 36
progression, 16, 18, 30
promotion, 16, 19, 23, 28, 30
  of carcinogenesis, 33, 36
promotor
  of carcinogenesis, 33
promutagen, 33
prostate cancer, 22
prostate carcinogenesis, 36
proteases, 36
putrefactive bacteria, 70
putrefactive metabolites, 72

quercetin, 23

radiation, 18
radical scavenging power, 95
rat bone marrow cells, as test system, 32
rats, 19
rejuvenaion, 18
repaire
  of initition, 33
reproduction, as cause of cancer, 15
resemary, 23
respiratory organs, 73
retinoblastoma, 16
reverse-transcriptase, 75, 76
riboflavin, 83
rice and peanut molds, 19
rice, as carcinogen, 20
room temperature
  for storage of tea, 59
Rosa, 24
rosemary, 23
γ -rays, as mutagen, 32
rverse transcription, 75
Ryokucha (Green tea), 9

safflowers, 83
saffron, 23
Salmonella typhimurium, 19, 30
  as test system, 32
salt fish, 20
salted samon filets, 82
salty foods, as carcinogen, 20
saponin
  contained in tea, 94
Sashimi, 99
school children, 63
semi-fermented tea extract
  suppresss histamine, 75
Sencha, 9, 29, 35, 75
  best temperature for, 89
  how to pour, 90
  suppresses histamine, 75
senile dementia, 56-58
senile spots (plaques), 56
Sen-no-Rikyu, 5
sepicemia, 73
sexual behavior
  as cause of cancer, 14, 15
sexual intercourse, 75
shellfish
  inhibits ACE enzyme, 42
Shen Nung, 1
short-term tests
  for carcinogenicity, 19
for mutagenicity, 28
SHR
  see spontaneously hypertensive rats
SHRSP: *see* stroke-prone SHR rats
sick house syndrome, 97
silkworm
  see Bombix mori
skin tumors
  inhibited by catechins, 32
sleepiness, 12

slug repellant spray, 97
small intestinal tumors
  inhibited by catechins, 32
smokers, 18
sneezes
  in allergic reactions, 73
snivels
  in allergic reactions, 73
soft drrink industry, 93
soft water, 89
soil, 13
soybeans, 19
  inhibit ACE enzyme, 42
spices, 23
spontaneous mutations, 29, 32
spontaneously hypertensive rats, 42, 43
sport drinks, 93
sprin water wells, 89
Staphilococcus, 70
  inhibited by tea catechins, 67
starch foods, as high risk foods, 20
starch, 48, 49
starchy foods, 20
Streptococcus mutants, 65
stomach cancer, 20-22
stomach cancer mortality, 25
stomach cancer patients
  in Shizuoka Prefecture, 25
storage of tea, 59
storage tea, 79
streptomycin, 13
stroke-prone SHR, 43
sucrase, 46
  inhibited by tea catechins, 49
sucrose, 48, 49
sugar-resolving enzymes, 45, 47
suitable water, 89
superoxide lipids, 81
superoxide, 81
Sushi, 99
sweetend black tea
  kills bacteria, 68
swelling of blood vessels, 73
synergistic effcts, 74

taboos in tea drinking, 91
Taiwan oolong, 75
tannic acid, as antimutagen, 32
taste abnormality, 12
Tea and Health Promotion, 3
tea caffeine
  stimulates neural transmitter production, 56
tea catechins
  decrease liver fats, 53
  diminishes garlic odor, 67
  have synergistic effects, 74
  improve mental activity, 56
  inhibit α-amylase activity, 46
  inhibit bacterial enzyme, 64
  inhibit bacterial growth, 67
  killes bad bacteria, 69
  kills MRSA, 73, 74
  promote good bacteria, 69
  supress blood sugar levels, 48
tea catechin-treated filters, 85
tea ceremony, 4, 5, 89
Tea Classic, 1
tea drink, 93
  in Shizuoka Prefecture, 27
tea instructors, 28
tea rice, 79
tea stalk, 80
teanine
  stimulates NGF synthesis, 57
tears
  in allergic rection, 73
teleocidine, as carcinogen 32
TF
  as anticarcinogen, 32
  as antimutagen, 32
  inhibits α-amylase activity, 46
  inhibits bacterial enzyme, 64
  see theaflavin
TF2, *see* theaflavin diggalate
TF2A
  as antimutagen, 32
  inhibits α-amylase activity, 46
  inhibits bacterial enzyme, 64
  see theaflavin mono gallate A
TF2B
  inhibits α-amylase activity, 46
  inibits bacterial enzyme, 64
  see theaflavin monogallate B
TF3
  as anticarcinogen, 32
  as antimutagen, 32
  has antibacterial effect, 62
  inhibits α-amylase activity, 46
  inhibits bacterial enzyme, 64
  kills MRSA, 73
  prevents influenza virus infection, 61
theaflavin, *see* TF, 29
  as anticarcinogen, 24
  chemical structure of, 31
  inhibits platelet aggregation,, 38
theaflavin digallate
  as anticarcinogen, 24
chemical structure of, 31
theaflavin gallate
  as anticarcinogen, 24

theaflavin monogallate A, as mutagen, 32
chemical structure of, 31
theaflavin monogallate B
chemical structure of, 31
theanine
three Kawanes (3K town), 26, 27
tobacco, 18
as cause of cancer, 14, 15
as mutagens, 18
Tradescantia, 19
transfusions, 75
Trp-P-1, 19, 20, 30
as mutagen, 32
Trp-P-2, 19, 30
tryptophan pyrolysate, see Trp-P-1
tryptophan pyrolysis products, 23, 30
tube formation, 37
tuberculosis, 13
tumeric, 23
Tungting oolong (Taiwan oolong), 75
twin sisters, 12

Ultraeye (TV program), 56
unfemented tea, 9
unsaturated fatty acids, 82
unsavory meat flavors, 84
Ura-Senke, 28
urinary passages, 73
urokinase, 34
used tea leaves, 79
UV, as mutagen, 32
as cause of cancer, 14
UV irradiation, 13, 30

vanilla, 23
vanillin, 23, 24
vascular endothelial growth factor, 37
vasocontraction of microcirculation, 42
vegetables, 20, 23
VEGF
see vascular endothelial growth factor
very low-density lipid cholesterol, 52
Vibrio, 67, 70
inhibited by tea catechins, 67
vinyl chloride, 19, 20
vitamin A, 82
as anticarcinogen, 34, 35
vitamin C, 23
as anticarcinogen, 24
as antimutagen, 35
as antioxidant, 82
contained in tea, 94
prevents color fading, 83
vitamin E, 23, 36
as antioxidant, 82
contained in tea, 94
prevents color fading, 83
vitamins, 23
VLDL
see very low-density lipid choresterol

warts, 76-78
water, 13
water-insoluble glucagon, 63
Wilms syndrome, 16

X-irradiation, 13
as cause of cancer, 14

Yabukita, 8, 75
yeast
for mutagenicity test, 28
yeast, as test system, 19
Yeisai, 3, 11, 56
yellow rice, 20
yogult
controls digestive tract, 72
Yoshimitsu Ashikaga, 5
Yunnan, 1, 2